Critical Approaches to Children's Literature

Series Editors
Kerry Mallan
Cultural & Language Studies
Queensland University of Technology
Brisbane, QLD, Australia

Clare Bradford
Deakin University
Burwood, VIC, Australia

This timely new series brings innovative perspectives to research on children's literature. It offers accessible but sophisticated accounts of contemporary critical approaches and applies them to the study of a diverse range of children's texts - literature, film and multimedia. Critical Approaches to Children's Literature includes monographs from both internationally recognised and emerging scholars. It demonstrates how new voices, new combinations of theories, and new shifts in the scholarship of literary and cultural studies illuminate the study of children's texts.

More information about this series at
http://www.palgrave.com/gp/series/14930

Anna Kérchy · Björn Sundmark
Editors

Translating and Transmediating Children's Literature

palgrave
macmillan

Editors
Anna Kérchy
University of Szeged
Szeged, Hungary

Björn Sundmark
Malmö University
Malmö, Sweden

Critical Approaches to Children's Literature
ISBN 978-3-030-52526-2 ISBN 978-3-030-52527-9 (eBook)
https://doi.org/10.1007/978-3-030-52527-9

Cover credit: Heirs of Olga Siemaszko

This Palgrave Macmillan imprint is published by the registered company Springer Nature Switzerland AG
The registered company address is: Gewerbestrasse 11, 6330 Cham, Switzerland

Preface

Writing, like translating or transmediating, is never a solitary activity. Academic collaborations act as major sources of inspiration for creative endeavors. The editors of this volume first met by courtesy of a vast international children's literature translation project organized to celebrate the sesquicentennial anniversary of the publication of *Alice's Adventures in Wonderland*. During this project, the scholars who worked on retranslating into English the Mad Tea Party chapter from 150 different languages—under the editorial supervision of Jon Lindseth and Alan Tannenbaum—had the unique chance to exchange ideas and build professional bonds as well as friendships grounded in shared enthusiasm. Researchers in the same field keep gaining collective impetus from sessions like the 2012 IBBY congress in London revolving around the theme of *Crossing Boundaries: Translations and Migrations*—where, in the keynotes, Patsy Aldana urged to give every child a voice by publishing in the dominant languages children's books from minority cultures, and Emer O'Sullivan argued that children's universal right to read should be satisfied by translating the best of children's literature from around the world into a plethora of possible languages. More recent events include The Child and the Book Conference in 2014; the interdisciplinary conference Children's Literature and Translation—Current Topics and Future Perspectives, co-organized in Fall 2017 by KU Leuven's Elke Brems and the University of Antwerp's Vanessa Joosen, featuring Jan Van Coillie's CERES lecture on how diversity can change the world; and the From

Morals to the Macabre in Translations for Children in Spring 2018 at Krakow Pedagogical University, organized by Joanna Dybiec-Gajer. In fact, some of the essays published in the present volume were originally presented at these meetings. However, the most immediate source of inspiration for the present volume was the impressive amount of submissions generated by a call for papers to a special issue of *Bookbird: A Journal of International Children's Literature* (2018/1) devoted to the theme of translating and transmediating.

Many contributors to the present volume had the chance to meet and discuss the translatability of silence and silencing in children's literature and its adaptations at the 2019 IRSCL Congress in Stockholm. We hope that our collection will continue to stimulate further discussions about the translation and transmediation of children's literatures and culture.

Szeged, Hungary — Anna Kérchy
Malmö, Sweden — Björn Sundmark

Contents

List of Contributors

Aneesh Barai University of Sheffield, Sheffield, UK

Clémentine Beauvais University of York, York, UK

Agnes Blümer ALEKI, University of Cologne, Cologne, Germany

Dana Cocargeanu Independent Researcher, Alumna of Dublin City University, Dublin, Ireland

Cheryl Cowdy York University, Toronto, ON, Canada

Joanna Dybiec-Gajer Pedagogical University of Krakow, Krakow, Poland

Hannah Felce Cardiff University, Cardiff, UK

Casey D. Gailey Vanderbilt University in Nashville, Nashville, Tennessee, USA

Anna Kérchy University of Szeged, Szeged, Hungary

Carl F. Miller Palm Beach Atlantic University, West Palm Beach, FL, USA

Petros Panaou University of Georgia, Athens, GA, USA

Karolina Rybicka-Tomala Jagiellonian University, Kraków, Poland

Annalisa Sezzi University of Modena and Reggio Emilia, Modena, Italy

Domingos Soares Universidade Federal de Santa Catarina, Florianópolis, Brazil

Cybelle Saffa Soares Universidade Federal de Santa Catarina, Florianópolis, Brazil

Björn Sundmark Malmö University, Malmö, Sweden

Tasoula Tsilimeni University of Thessaly, Volos, Greece

Dafna Zur Stanford University, Stanford, CA, USA

List of Figures

Translated Book Covers as Peritextual Thresholds: Comparing Covers of Greek Translations to Covers of Source Texts

Translating Tenniel: Discovering the Traces of Tenniel's Wonderland in Olga Siemaszko's Vision of *Alice's Adventures in Wonderland*

Between Light and Dark: Brazilian Translations of Linguistically Marked Ethical Issues in Star Wars Transmedia Narratives for Children

Translating Ambiguity: The Translation of Dual Address in Children's Fantasy During the 1950s and 1960s

Introduction

Björn Sundmark and Anna Kérchy

Translation and transmediation—the telling of a story across media—are related practices. Translation regularly involves transmediation, which in turn can be seen as a specific form of translation. In this collection we explore how translation and transmediation plays out in children's literature across languages, genres, and media forms. Multimedia literacy is the foundation of transmedia storytelling: the telling of story across a variety of media platforms, formats, and techniques and the creation of an original yet adaptable story world. It also creates multiple points of entry for the listener of the story that is being (re)told, and invites audiences to play an active role in putting together the puzzle pieces of meaning (Jenkins). As a collective narrative practice, in which parallel storylines, episodes, and backstory may be accessed through different media, transmedia storytelling ties in with Linda Hutcheon's definition of adaptation as a process of creation and reception in which new, autonomous texts are constructed through transformative processes of retellings, revisions, and repurposings. While an adaptation's acknowledgment of the original may not

B. Sundmark (✉)
Malmö University, Malmö, Sweden

A. Kérchy
University of Szeged, Szeged, Hungary

A. Kérchy and B. Sundmark (eds.), *Translating and Transmediating Children's Literature*, Critical Approaches to Children's Literature,
https://doi.org/10.1007/978-3-030-52527-9_1

be obvious (fans of Disney's Little Mermaid are not necessarily familiar with Andersen's tale), transmedia narratives tend to foreground their "extended intertextual engagement with the adapted work" (Hutcheon 35) while making a distinctive and valuable contribution to the familiar yet refashioned storyworld.

Transmedia storytelling can be spontaneous and unsynchronized, as in fan responses to a narrative, but it can also be employed as a conscious, synchronized marketing strategy. It should also be noted that whether transmediation is part of the story's conceptualization and original form of production or a process drawn out over time—as with *Alice in Wonderland* or *The Hobbit*—is of little critical importance. From a reader/audience perspective, transmediation essentially means that there is more to the story than meets the reading eye. Thus, the story takes on an existence across media and cannot be reduced to a single authoritative source text, nor can it simply be assigned a "natural" format or language. Transmediation allows audiences to "claim aspects of ownership over content that they can identify with, immerse themselves in, adapt, remix, reuse and share," in analogue and digital realms, driven by "variation and repetition," "porousness, instability, collaboration, and participation on a global scale" (Hutcheon and O'Flynn 206). Transmedia storytelling is all about piecing together, adding to, embellishing, editing, transforming, and translating.

Translation—the second key concept of this research anthology—is akin to that of transmedia storytelling in so far as both involve a transition between semiotic systems. Although a translated verbal text does not require any other medium than that of the first text, translation, like transmediation, involves reinterpretation, and it provides new versions of a source text (or several). Moreover, for a bilingual reader, the experience of reading both the source text and its translation is similar to the experience of transmedia storytelling in that the translation provides an alternative or complementary version to the source text. As with transmedia storytelling, translation, too, provides alternative narrative points of entry. The first text version that a reader encounters is not even necessarily the source text; you can start with the translation and then go on to read the source text in the original language. Just like transmedia storytelling is a participatory cultural performance of networked fan communities of rewriters and rereaders, translation is necessarily a collaborative act, that involves a network of actors including authors, translators, editors, dubbing adapters, publishers, and to a certain extent audiences, too, who "share authority over the work or at least some portions of it" (Cordingley and Manning). Translation like transmediation is a "creative and interpretive act of appropriation/salvaging" (Hutcheon 35): an old text is rescued from oblivion by being turned into a new text.

Educational psychology recognized the kinship of the transmediation and translation process as early as in the mid-1980s, contending that learners with limited linguistic capacities can be encouraged first to perform in the musical, spatial, and bodily kinesthetic realms before attempting to represent the parallel meaning through a foreign language. Charles Suhor defined the term transmediation as "the student's translation of content from one sign system into another." He urged teachers to open up their classrooms for the use of many "signways" by integrating into their syllabus "literal transmediation" ("making a raft like the one described in *Huckleberry Finn*; writing a paraphrase of a poem read in class, making a slide-show to illustrate a short story; or doing a mime that parallels the action in a narrative poem") and "imaginative transmediation" ("a collage based on a book; a critical review of a film; a role-playing exercise based on the theme of a story; or a free writing exercise in response to [a picture or] an instrumental musical recording" (Smith 191)). By the postmillennial era, these intersemiotic ventures have evolved from innovative pedagogical tools into inevitable strategies of socio-cultural negotiation.

When Susan Bassnett and André Lefevere announced "the translation turn in cultural studies" in 1998, they prognosticated a major paradigm shift of the postmillennial era throughout which translation would provide a potent metaphor and efficient analytical framework to deal with socio-political transformations and upheavals such as globalization, the post-9/11 crisis of multiculturalism, or migration—all concomitant with discursive conflicts necessitating cross-cultural as well as cross-generational negotiations. Since then, the interdisciplinary research of children's literature and cultures has gained a considerable impetus from translation studies' strategies designed to balance the hegemonic power play involved in textual and social exchanges. Seminal works—such as Riitta Oittinen's *Translating for Children* (2000), Emer O'Sullivan's *Kinderliterarische Komparastik* (2001), Jan Van Coillie and Walter P. Verschueren's *Children's Literature in Translation* (2006), or Gillian Lathey's *The Role of Translators in Children's Literature: Invisible Storytellers* (2010)—reveal how the formerly underestimated art of children's literature in translation may eventually open doors for future generations toward adventurous, empathic explorations of cultural differences and shared communal delight in finding consensual meanings grounded in transnational understanding, solidarity, trust, and imagination.

In fact, the translation turn ties in with "the digital turn" of the twenty-first century, in which an ever-growing flood of digital information technologies radically transforms our understanding of the human world, and builds bridges between old and new media, material and virtual reality, computer and human intelligence (see Westera). Thus, conjoining the methodological apparatuses of new media/adaptation studies and translation studies with those of children's and young adult literature criticism seems inevitable in a technologically enhanced epoch when young "prosumers"[1] (Manovich 3) of cultural products grow up as native speakers of the digital language of computers, smartphone applications, online social media platforms, video games, and downloadable e-books.

In today's "digimodernist" (Kirby) post-industrialist consumer societies, mass-marketed technological devices have radically altered children's and young adults' reading experience. Generation Z (the demographic cohort following the Millenials) and, even more so, the succeeding Generation Alpha (anyone born after 2010) are digital natives and natural-born "produsers" (Marshall 2004) with an unprecedented new media literacy and online interactive telepresence; they have been trained to understand the art of storytelling in terms of a multimodal process involving "dispersed media content" (Jenkins 3). A significant economic factor and consumer group increasingly independent of parental supervision (Beeler and Beeler 2), twenty-first-century youngsters no longer wait for a family outing to the local library or cinema but can easily and simultaneously download in the comfort of their own rooms e-books, iPad apps, films, fan videos, or DIY tutorials revisiting the same fictional universe, often in a variety of different languages. They spontaneously make intertextual connections over multiple media surfaces while seeking out new information about the artwork that lends itself to creative interactions with customizable special features (medial extensions like Snapchat filter applications, Facebook quizzes, or added "playable" bonus contents, like the ones at Disney Movies Online or J. K. Rowling's *Pottermore*) and that incites further imaginative reinventions to the producers' own liking. They take for granted that any fantasy realm can be enhanced by fanart, videoblog commentaries, or cosplay performances—which are destined to be shared on online social media with hashtags guaranteeing the global unification of the multisensorial, transmedial narrative experience.

Linguistic enhancement is an integral part of the transmedia extension process as cultural products shared on the internet can be complemented by subtitles, dubbings, comment streams in any foreign language of one's choice. This polyphonization, the complementary or contradictory coexistence of the source- and sub-voices, will increase the transnational appeal of the original, as the immense online popularity of the multilingual performances of Disney animations' theme songs have attested. (The video compilation of *Frozen*'s "Let it to go" sung in twenty-five different languages, starting with English and seamlessly transitioning from one language to the next, had over seventy million viewers on YouTube since 2013 and earned critics' praise for "allowing us to appreciate the audible beauty of other languages with a familiar song").

Both transmediation and translation are "creative derivative methods," to use Venuti's words ("Adaptation, Translation, Critique" 29). These are "methods" that reenact the meaning of a de/recontextualized source text that is inevitably modified in its content and form while remaining a dialogic reference point. Jay David Bolter and Richard Grusin's pioneering 1999 book *Remediation* outlined a theory of mediation for our digital age, arguing that new visual media achieve their cultural significance precisely by "remediation," by paying homage to, rivaling, and refashioning such earlier media as perspective painting, photography, film, and television, while alternately attempting to ignore or deny the presence of the medium and/or foregrounding a fascination with the medium. The translation process is driven by similar, simultaneously complementary and contradictory aims at achieving "transparent immediacy" and "hypermediacy" (Bolter and Grusin): the struggle to stay true to the original and to make one's own voice and language heard, to respectfully pay homage and to transfigure in a necessarily iconoclastic way.

Mediation—whether in the form of adaptation, translation, or remediation—allows for the reevaluation of a variety of literary theoretical concepts ranging from authenticity, textuality, authorship, audience agency, age appropriateness, storytelling, or imaginativeness, while foregrounding the ideological interests, the educational and ethical responsibilities, and the semiological complexities involved in the transition process. The most pertinent questions can be the simplest, like the one put forward by Michael Cronin in his *Translation in the Digital Age*: "If translation has typically depended on a deep commitment to, and reverence for, the printed word, what happens when the experience of the printed word shifts from the page to the screen?" (6).

Henry Jenkins' 2007 definition of "transmedia storytelling"—a systematic dispersal of integral elements of a storyworld across multiple delivery channels which each make a unique, original contribution to a coordinated entertainment experience—encapsulates the worldbuilding strategies of most of today's popular children's literary/cultural products. The lure of Alice in Wonderland, the Little Prince, Harry Potter, Aladdin, Chihiro, or the Moomins is considerably enhanced by the plethora of interconnected media platforms (novel, film, animation, computer game, fanfiction, cosplay, collectibles, etc.), all of which maximize audience engagement by unfolding an increasingly elaborate fictional reality. The way in which each media "adds a new cultural layer, supporting more diverse ways of communicating, thinking, feeling, and creating than existed before" (Clinton, Jenkins, and McWilliams 11) resonates with how translation as an inventive "act of both inter-cultural and inter-temporal communication" (Bassnett, *Translation Studies* 9) allows us to see in different ways the original text that always already "bears in itself all possible translations and gets richer with each additional reading-rewriting," as Walter Benjamin puts it (17).

Transmedia storytelling has proved to be one of the most influential notions for the critical analysis of postmillenial cultural production. Nevertheless, although iGenerations certainly excel in maximizing their engagement with the storyworld through individual choices, additions, and creative retellings and have learned to negotiate media transition as a "mix of tradition and innovation" (Jenkins and Thorburn), the multidimensional expansion of a storyworld involved in transmedia storytelling is not a uniquely post-postmodern, twenty-first-century phenomenon. Already in the 1980s, long before tablets and mobile apps, transmedia storytelling endeavors targeted child readers with *enhanced texts* like the Care Bear book series (launched by the board game company Parker Brothers) that came with a cassette tape for sing-along songs, a VHS video to watch the animated cartoon version of the narrative, and a stuffed bear toy to cuddle and was developed as a franchise marketed in many different languages from the very beginnings (e.g., French, German, and Hebrew). Even the movable, three-dimensional paperart of pop-up books—emerging as early as the fourteenth-century fold-out anatomy books for medical students and gaining increasing popularity in the 1700s with the "lift-the-flap" book design targeting solely child audiences—offers an analogue precursor to the iPad apps' book-enhancing technologies.

We can wonder if there is a genuine difference between the reader making Wonderland's White Rabbit tremble with fear by shaking the picturebook *Nursery Alice* (1890) at the authorial instruction embedded in the print and paper edition for pre-readers and activating the same mobilization of the illustration, the momentary animation of the fictional character, by tapping on the digital device's screen? (Kérchy 33). The reader/listener's kinetic interpretive gesture bringing the fictional figure into being cannot help but remind us of the performative aspect of translation as an intercultural and inherently intersemiotic communicative act. Like the child shaking the book to animate the bunny, the translator "does things with words" in an Austinian sense to adjust linguistic representations to mental imaging. The translation analogy can be taken further if we consider the "interpictorial" (Kokkola) transition and explore how Gertrude Thomson's cover image, created specifically for *Nursery Alice*'s first print run, offers a revision of Tenniel's classic illustrations which constitute an integral part of Carroll's prose narrative to enhance the "Wonderland-experience" granted by the storybook. We actually have literary-historical evidence to support the clever conjoining of the complex dynamics of transmediation and translation: Carroll's correspondence[2] testifies that it was the very first Dutch translation of his *Alice's Adventures in Wonderland* decorated by "pictures done large in colours" that brought him the idea to adapt his novel in an abridged form for "children from naught to five." The Preface's dedicatory words about "the illiterate, ungrammatical, dimpled darlings" who will more likely "coo over" than actually read the dog-eared pages attests that a change in the age of the target audience will also inevitably elicit a transgenerational translation of the source text. The Carrollian meaning modification modes above map the issues that are the focus of our attention in the present volume: inter- and intra-cultural transformations, media transitions, iconotextual interactions, metapictorial potentialities, and intergenerational transmissions, which interact throughout the complex conjoint enterprise of adapting, translating, and transmediating children's literature.

Looking at it transhistorically, transmedia storytelling can emerge as a side effect of the narrative strategy of imaginary worldbuilding. The construction of an elaborate fictional universe often requires the mapping out of different details (characters, motifs, additional episodes, parallel scenarios) of an alternate reality on different generic- and media platforms which may inform, complement, or challenge one another—like Tolkien's

Middle-earth is enhanced by fictitious languages, legends, geography, historiography, and cosmology, which all have expanded into various genres and media platforms, crossing over into card-, board-, video-, and mobile games, films, cartoons, internet memes, and merchandise. This "adaptogenic" quality (Groensteen in Hutcheon 15) might be a reason for the enduring popularity of children's literary classics. After all, J. M. Barrie's Peter Pan travels from Kensington Gardens to Neverland through the intermediary of an "adult novel" chapter, and then a stage play, before reaching its iconic status as a children's novel; Peter Rabbit's rural realism can be tracked back to Beatrix Potter's picture letters, and Milne's Hundred Acre Woods is mapped out in a toy-focused story that lends itself easily both to the serious philosophical agenda of object-oriented ontology ("thing studies") and the light entertainment by the transmedia commodification of fantasy (over 10,000 Pooh Bear products available on Amazon).

Our starting points for the organization of the volume were very basic assumptions: The translation and transmediation of literature are related and interconnected practices. Translation is fundamental to literature, not least to children's literature. Without books in translation, we are locked inside our own literary ghettos. This is true even of large languages, like English, where the percentage of translated books is often very low. Transmediation is another revitalizing rewriting process by which works of literature are adapted to new purposes, media forms, and genres. Through such adaptation work, books are given new lease of life and can continue to promote understanding of different times, cultures, and languages.

However, only recent endeavors have started to consider translation and transmediation as reciprocal and complementary to each other (C. O'Sullivan, Sundmark, Kérchy, and Waller). Previous studies have mostly focused on the challenges of translation (Van Coillie), cultural transmission, didacticism, linguistic challenges (E. O'Sullivan; Oittinen), the role of the translator (Lathey), and the impact of specific children's books in translation (Beckett and Nikolajeva). In this collection we aim to contribute to the solidification of an emerging new research field by exploring the connections between translation *and* transmediation, covering a broad scope in terms of languages, dialects, and intermedial aspects.

The volume also enters into dialogue with the current critical discussion of adaptation in children's literature taking place in collections like

Textual Transformations in Children's Literature: Adaptations, Translations, Reconsiderations edited by Benjamin Lefebvre (2013), *Adapting Canonical Texts in Children's Literature* edited by Anja Müller (2013), and *Never-ending Stories: Adaptation, Canonisation and Ideology in Children's Literature* edited by Sylvie Geerts and Sara Van den Bossche (2014). Benjamin Lefebvre's cutting-edge book analyzes primarily adaptations, abridgments, parodies, and mash-ups in the field of children's literatures and cultures; and mostly uses the word *translation* in a metaphorical sense, as a synonym for media shift. Yet two of the dozen chapters also deal with children's literature in actual translation (one discussing Vietnamese folktales and the other Polish adaptations of Perrault), hence making the first steps in the direction that is the focal point of our present volume. *Never-ending Stories*—notable for its global breadth, cross-cultural comparisons, and a broad range of genres tackled, from oral tales to digital magazine fiction—identifies, in its introduction, translation as a form of adaptation which widens the distribution of a pre-text, and adaptation as a translational strategy determined by the translator's ideological assumptions and conceptions of childhood. Transmediality comes into the picture as one of the three fundamental aspects of adaptation (besides the socio-political and socio-cultural aspects), but the overall agenda of the volume is to bridge the gap between adaptation and canonization studies, and the study of children's literature—just like in the case of Müller's book. Our project gained further inspiration from collections which explore the dynamics of the international reception of children's classic tales driven by the agenda to explore the verbal (and visual) manifestations of transculturally constant and locally particular features across many languages and cultures, such as *Grimms' Tales Around the Globe* edited by Vanessa Joosen and Gillian Lathey, and *Cinderella Across Cultures* edited by Martine Hennard Dutheil de la Rochere, Gillian Lathey, and Monika Wozniak.

The aim of this collection is the timely endeavor to approach translation and transmediation as an interrelated practice and apply it to the analysis of children's literature. The project is meant to go well beyond the default British-American context and look into translation from and into neglected languages and dialects. Tellingly, the authors of the seventeen chapters come from different countries and deal with translations into/from a variety of languages (including Brazilian, French, German, Italian, Swedish, Swiss, Hungarian, Polish, Romanian, Serbian, Korean, Greek, and Latin). Fourteen of the sixteen chapters selected for this present volume deal with language change and verbal meaning transition in connection with cultural transition and transmediation. Moreover, the project brings up practices regularly overlooked in transmediation/translation studies—such as trans-sensory new media storytelling, adapting picturebook covers, and the toddlerization of science.

The chapters deal with an exciting variety of topics. Some study issues of globalization/localization/glocalization, ideological shifts, and ethical agendas involved in "domesticating and foreignizing" translation strategies (Venuti, *The Translator's Invisibility*), reconceptualizations of fictional elsewheres and reimagined homes, cultural sameness and difference through media- or language change. Others tackle the responsibility of mediators in (re)constructing the image/voice of the child reader and the translation/transmediation of children's and young adult literature as a negotiation process between publisher demands, parental expectations, social norms, children's cognitive abilities, emotional needs, and fantasizing agency. A few chapters discuss how intergenerational, intergeneric dynamics fuel crossover fiction's dual audience engagement; others focus on the functioning of "image-textual dynamics" (Mitchell 89), the relation of verbal and visual representation across a variety of media, illustration as intersemiotic translation, and the challenges "the narrative art of picture books" (Nodelman) poses for translators. The articles' common denominator is an emphasis on how translating and transmediating children's literature enhances the genre's unique potential of an "education by fantasy." The proliferation of versions, adaptations, retellings, and revisionings of the same story (world) across a variety of languages, media platforms, and communication channels foregrounds how the imaginative construction of non-existent but possible worlds opens up political vistas by urging the empathic consideration of others' perspectives differing from our own and the recognition of our collective memory's role in shaping our understanding of past and future. Scholars of children's literature translation/transmediation studies seem to be particularly sensitive to how this "imaginative responsibility of confronting the world as we know it or as it might be or even as it might have been" allows "multiple ways of knowing: curiosity, creativity, pleasure, and imagination as the bedrock of reason in its most exalted form" (Wu, Mallan, and McGillis xi).

The present volume is divided into five parts, each of which sheds light on a vital aspect of translating and transmediating children's literature. The part headings are as follows: "Inter-/Intra-Cultural Transformations," "Image-textual Interactions," "Metapictorial Potentialities," "Digital Media Transitions," and "Intergenerational Transmissions." The contributions shed light on the translation of children's literature as a creative crosscultural transfer of artistic products. What these transmediations have in common is that they are designed for young audiences and that they excel in enriching the complex layers of the signification of the creative source text by wedding multimodal, crosslinguistic, iconotextual, metanarrative/-pictorial, and intergenerational dynamics. The topics highlighted in the section headings necessarily overlap with one another, and some (like "intracultural transformations" and "intergenerational transmissions") could be applied to all children's literature implanted from one cultural context to another by adults for children. Individual chapters will illustrate similarities, differences, and especially interrelations between these various aspects of translation and transmediation within children's literature, tracking their evolution over time, across regions, and through increasingly interdisciplinary disciplines. The fuzziness of the categories organizing the sections and the multifocal perspective of individual articles that tackle multiple section topics results in the flexibility of the volume's structure. It allows readers to choose their own order of reading, to skip from one section to another and then return, like in the case of anthologies and hypertexts, and hence become involved in the creative act of compilation. Readers are invited to become translators engaged in a dynamic process of recreating the text (created by the authors, and editors). In line with Hans Georg Gadamer, we believe that "Reading is already translation, and translation is translation for the second time…The process of translating comprises in its essence the whole secret of human understanding of the world and of social communication" (Schulte 2).

Part one explores the inter- and intracultural transformations that are integral to the translation-transmediation process. The process of translating is never simply a matter of rendering meaning (semantic information) from one language into another. It always also implies *mediating* from one culture into another (cultural codes/signification), bridging different perceptions of the world, ways of thinking, social roles, and background knowledge, to reconcile differences that could act as communication barriers. Contributors study tensions and transitions between lingua franca and minority languages, regional dialects, ideologically influenced and politically correct discourses.

Clémentine Beauvais's chapter offers a diagnosis of the current status of European children's literature in translation in the United Kingdom today, in the age of Brexit. Her multifocal analysis ties the kaleidoscopic concept of cultural, individual, and linguistic *difference* to the enriching cultural alterity and diversity translated texts should expose young readers to, as well as to the special methodological tools the study of children's literature in translation on the British isles requires in a time of unprecedented political tension with "the Continent." Beauvais introduces Francois Jullien's "exploratory concept" of the *écart*—the gap or sidestep—as a key term that may offer a conceptual opening toward a more flexible theoretical and ideological framework for intercultural analysis, more successfully tuned into the "committed aesthetics" of children's literature in translation in the United Kingdom. The opening of *Toby Alone* (2008), in Sarah Ardizzone's translation of Timothée de Fombelle's *Tobie Lolness* (2007), offers an example of a "translation's elasticity in that sensitive zone." Beauvais's conclusion, filled with hope against all odds, argues that translating children's literature might eventually function as a political gesture, opening up a bypass guiding out of ethnocentric shortsightedness, out of ideologically prescribed dead-ends, allowing readers to wonder and wander in-between cultures and languages.

Hannah Felce's case study focuses on intra-cultural transformations of a Swiss national classic, Selina Chönz's Alpine tale *Uorsin*—originally written in one of the dialects of Switzerland's fourth national language, Romansh, called Ladin, and illustrated in a picturebook format by Andersen Award winning Alois Carigiet. (It was translated into English as *A Bell for Ursli: A Story from the Engadine in Switzerland* and recently adapted into a family adventure film under the title *Schellen-Ursli* [2015].) Felce's chapter demonstrates through the example of *Uorsin* how the publication of a children's picturebook written in a minority language calls into question the hierarchical distinction and chronological sequentiality presumed between a predetermined original and its secondary translation(s). The prioritization of the written source over the complementary illustrative image can be further problematized by *Uorsin*'s two sequels, *Flurina und das Wildvöglein* (*Florina and the Wild Bird*) and *Der grosse Schnee* (*The Snowstorm*) which were first written in German as accompaniments to the Sursilvan-speaking Carigiet's original illustrations. Felce's focus on the multilingual publishing procedure's specificities, the intracultural transition, and the "intralingual textual transformation" (a term she uses by conjoining Jakobson's and Lefebvre's concepts), as well as on "intersemiotic translation as adaptation" allows her to challenge the binary model of the translation process and of the fixed nature of language.

Joanna Dybiec-Gajer discusses the intercultural transformations involved in the eastbound journey of Heinrich Hoffmann's children's classic *Der Struwwelpeter* (1845). A short overview of Polish-language editions explores the changing context of the translation, tracing the publication history of the book enhanced by numerous spin-offs or struwwelpetriades while mapping the evolution of the notion of invisible translator into a celebrity translator. Relying on Emer O'Sullivan's narratological approach to translation, Dybiec-Gajer studies how the modifications of the narrator's voice in subsequent rewritings affect the relation with the source text's original narrator-focalizer and the implied child reader. The chapter argues that the distance and reticence of the German *Struwwelpeter*'s narrator, have posed a considerable challenge for the mediators of the book. A close-reading of Polish-language editions reveals that non fail to make the narrator's voice "more audible" than in the original, with the aim to convert the uncoventionally ambivalent, absurd, and abject text into a more familiar didactic exercise. An analysis of the image-textual dynamics taking a variety of forms in the many retellings and revisionings of this picturebook reveals further aspects of intercultural transition.

Dafna Zur explores the neglected territory of politically informed intercultural transition in translating and transmediating children's literature. She argues that the most popular Soviet-inspired cultural forms emerging in the 1950s in North Korea were the children's literary subgenres of travel writing and science fiction; they offered a translation of the world in both a literal and figurative sense and provided a compass for the country's political and cultural visions. Zur's cultural-historical analysis explains how the 1950s in North Korea were marked by an effort to forge a new society out of the ruins of the Korean War, and how the Soviet Union and its children's literary products played a leading role in this process. Travel writing was punctuated by North Koreans' admiration of Soviet children's culture as a way of calling attention to the work that needed to be done at home; at the same time, they offered opportunities to assert budding

North Korean pride. As for science fiction, it was introduced as a new genre through translations from Russian, which brought together scientific education and a socialist moral vision. However, far from representing mere "imitations" of Soviet models, the translations reveal a complex negotiation of ideas about gender and science at a critical moment in North Korea's early literary development.

Part two studies "image-textual interactions" with the aim to explore the synergy of words and pictures in translation in (re)illustrated children's literature. In line with W. J. T. Mitchell's picture theory, the hyphenated title "image-text" refers to our interest in "*relations* of the visual and verbal," but contributions also tackle "image/text" as "a problematic gap, cleavage, or rupture in representation," and "imagetext" as "a composite, synthetic work (or concept) that combines image and text" (89), literally manifested in the picturebook form itself.

Aneesh Barai focuses on "foreignizing domestications and drawing bridges" in one of James Joyce's only known pair of stories written for children—a trilingual letter sent to his grandson in 1936 that was turned three decades later into *The Cat and the Devil*, a picturebook illustrated and translated into thirteen languages.[3] Joyce's postscript playfully calls attention to the fusion of polyglotism and neologisms, pivotal narratological devices of modernist novels, that he toys with on crafting a fable abundant in metalinguistic commentaries. Barai shows how Joyce domesticates a local folk legend of Beaugency by inoculating it with Irish wit to best engage his child reader, and how he "overlaps cats and letters in their textual functioning as bridges between cultures." We learn how translators have struggled to reproduce the Joycean effect of the source text by "foreignizing what was originally domestic: French language itself." Barai also considers the role of illustrations in translating verbal tricks to visual realms, with tongue-in cheek twists of pictures "enhancing" (Nikolajeva and Scott 8) the text: Rose's illustrations cast Joyce himself in the role of the devil, who is tricked by the manipulation of meanings, as he agrees to build a bridge for the city overnight but ends up cheated of his payment of the first soul to cross the bridge when a cat crosses it first.

Björn Sundmark offers an analysis of the translation and visualization of Tolkien's *The Hobbit* into Swedish, the aesthetics of fantasy, and Tove Jansson's illustrations. J. R. R. Tolkien's *The Hobbit* (1937) and *The Lord of the Rings* (1954) are commonly seen as instrumental in establishing fantasy as a publishing genre. However, at the time of publication, there were no established models or conventions for how (or even if) fantasy should be illustrated, and Tolkien's writings on the aesthetics of fantasy, as well as his own illustrations, have inspired subsequent illustrators but also led to a visual orthodoxy on how Middle-Earth and its inhabitants should be portrayed. In this chapter, Sundmark argues that by looking at the early translations and transmediation of Tolkien's work we can get a glimpse of alternative ways in which Tolkien's work can be (and was) interpreted. It is also shown that today there is a new openness to unorthodox ways of visualizing fantasy and a growing acceptance of the pioneer illustrators of the 1960s. The chapter focuses in particular on the expressive and non-realistic artistry of the work of Tove Jansson for the 1962 Swedish edition of *The Hobbit*. Ultimately, the chapter makes a plea for a reassessment of Jansson's Hobbit-illustrations based on the visual diversity evident in much present-day fantasy.

Anna Kérchy scrutinizes the (un)translatability of literary nonsense designed for children through a case study that tackles the iconotextual monstrosity of Lewis Carroll's "Jabberwocky" and its Hungarian domestications. The chapter studies how the interaction between verbal narrative and visual illustration, oral performance and written transcription, source text and translation can be regarded as vital constituents of the complex signification process Henry Jenkins calls "transmedia storytelling." Pictorial, acoustic, and crosslinguistic interpretations enhance the source-textual meanings in many ways, mutually formative of one another, each augmenting new dimensions of the immersive and coordinated entertainment experience. Kérchy explores the impossibility of translating nonsense on three levels. First, the chapter comments on "Jabberwocky"'s metafantastic, metanarrative interpellation of the reader as a translator who will inevitably both decode and reproduce nonsense—both on the level of implied readership (as in fictional Alice's quest for meaning) and through the actual lived interpretive experience of the tale's putative reader or listener. Second, the focus falls on illustration as a translation of the written narrative and the author's, translator's, and illustrator's self-conscious use of iconotextual dynamics. Kérchy argues that the embedded visual depiction of the Jabberwock (multiplied via the mirror-written *picture*-poem and Tenniel's original drawing) function by means of transmedial addendums, visual translations of verbal nonsense, meant to assist the child reader in a non-didactic, ludic manner to face the interpretive rite of passage involved in the attempt to tame textual monstrosity. Finally, the chapter briefly introduces the cultural transposition, linguistic transfer, and creative individual solutions emerging in domesticating Hungarian verbal and visual translations of "Jabberwocky."

The next section, "metapictorial potentialities," studies how pictures can be translated into other pictures while providing a self-reflective commentary on limits and potentials of visual representability itself. The dialogue between images—as in the case of the re-illustration of the same story—reflects on the intersections of visuality, language, similitude, iterability, and difference. Metapictoriality elicits a multifocal perception of verbal and visual experience, interrogating the authority of language over image, of original over transmedia adaptation.

The first chapter under this heading is a contribution by Petros Panaou and Tasoula Tsilimeni, where they compare source-text book covers with those of their Greek translations, drawing conclusions from a rich corpus of sixty-eight children's and young adult books published by major Greek children's literature publishers, Patakis and Psychogios, over the past ten years. According to Panaou and Tsilimeni, imported children's literature has had an important place in Greece since the nineteenth century, and translations (mainly coming from Western countries) have significantly influenced Greek authors' styles, themes, and techniques; nevertheless, comparative or translation studies focusing on the region's cultural, linguistic, and iconographic specificites are scarce. The chapter fills this gap through a "contact and transfer study" that focuses on the exchanges between literatures from different countries, languages, and cultures (E. O'Sullivan *Comparative Children's Literarure*) with the aim to explore how cover images, titles, designs, and other peritextual elements are translated, altered, adapted, omitted, added, or replaced. These interpictoral dynamics inform us about the transfer and reception of Greek translations for youth, as well as the trends, preferences, and sensitivities in Greek children's and YA literature in general. The chapter relies on Tsilimeni and Panaou's earlier research on the main factors influencing Greek publishers' decision-making when it comes to importing international texts, as well as Serafini, Kachorsky, and Goff's Multimodal Ensemble Analytical Instrument (MEAI), which is a tool designed specifically to guide critical multimodal analysis of book covers.

Karolina Rybicka investigates how the republications and retranslations of children's literary classics resonate with their re-illustrations, and she goes in search of illustrators whose pictorial translation of a literary text has proven to be considerably influential of the creative vision of all subsequent translators and illustrators of the same book. The chapter centers on the lasting impact of John Tenniel's 1865 vision of *Alice's Adventures in Wonderland*. Three types of "author traces" are distinguished to show how the Tennielian imagery informs different illustrations which, nevertheless, each adapt Wonderland into the visual, rhetorical style of their own target culture. The process becomes even more interesting when an interlingual translation is also a part of the equation. Rybicka offers a case study of Polish illustrator Olga Siemaszko's four versions of Alice and her Wonderland produced over thirty years to decorate different editions, each time translating Tenniel into her distinct artistic styles of the given period. Siemaszko created two strikingly different sets of illustrations for the Polish translation of Carroll's novel by Antoni Marianowicz (1955, 1969)—each time using the strategy of domestication present in the interlingual translation. Her other versions of *Alice* enhance Wonderland by transmedia extensions including a series of postcards (1964) and a vinyl cover for a 1975 radio-play adaptation of the story.

Part four, "Digital Media Transitions," tackles such pertinent issues as the translation of print and paper books into iPad apps and how digitalization affects the reading experience of young readers; the impact of transmediation on online translations of children's literature surfacing in paratexts, altered visual elements, or shared translation strategies; and difficulties of translating a transmedia storyworld franchise with a transnational appeal that has been dispersed and extended on a variety of media platforms from comics to computer games.

Cheryl Cowdy's "Grammars of New Media: Interactive Trans-Sensory Storytelling and Empathic Reading Praxis in Jessica Anthony's and Rodrigo Corral's *Chopsticks*" tackles the collective anxieties concerning how new media technologies, and the move from printed books to iPad apps, might debilitate the human capacity of empathy and replace "serious" reading experience with superficial entertainment of "uninvolved audiences." Her ethnographic reader-response methodology enters into conversation with actual child audiences to explore how curiosity and compassion may function as major engines of their interpretive activity, or how what Judith Fetterley would call a "resisting reading" can gain a therapeutical potential by allowing young readers to express their anger and discontent provoked by unspeakable traumas, such as the sudden loss of a parent. Cowdy also deals with the adult reader's empathic and critical responses to child readers' responses to the text and, hence, raises important questions about the cultural construction of childhood identified with vulnerable innocence. The research findings seem to suggest that the dual readership addressed by works like *Chopsticks* may also encourage a dialogic co-reading, a mutually enlightening intergenerational communication that critics of new media feared would disappear with the new technology.

The wedding of transmedia studies and affective narratology enriches the field of children's/YA literature scholarship—in line with recent cutting-edge projects as Moruzi, Smith, and Bullen's collection *Affect, Emotion, and Children's Literature* (2017)—and Cowdy's notion of "trans-sensory storytelling" will certainly stimulate further research in new media, trauma studies. It is also illuminating to consider how "the trans-sensorial narrative strategies operat[ing] as a kind of defense against the disappearance of the book and the body through a dual process of sensorial and technological extension" recycle analogue children's literary classics' "conventionally subversive" textual strategies. The autofictional patchwork identity in *Chopsticks*' iPad app (and the readers' affective/sensorial responses to it) are reminiscent of Anne Frank's diary, a trauma narrative in which the young autobiographer "writes into fleshly being" an embodied self she could not own in reality (Bishop 13) because she had to control all her corporeal urges while in hiding during the Second World War.[4]

Dana Cocargeanu discusses a case of double mediation of children's books, namely, the online translations of Beatrix Potter's tales in Romania. In the last decade, the print Romanian editions of Potter's stories have been supplemented by online translations posted by interested individuals. The chapter explores the factors that may have facilitated this phenomenon, the impact of transmediation on the translated tales, and the relationship between the online and the print translations of Potter's tales. Cocargeanu argues that the online translations were posted due to a growing interest in Potter, facilitated by the political, social, and cultural changes in post-Communist Romania; a scarcity of print translations of Potter's works; and increasing internet use among Romanians. The impact of transmediation on the translations is visible particularly in their paratexts and altered visual elements. However, the online and the print translations are also connected through shared translation strategies, such as the use of the "popular" register of Romanian to translate Potter's mostly Standard English.

Cybelle Saffa Soares and Domingos Soares focus on Brazilian translations of linguistically marked ethical issues in the Star Wars transmedia storyworld franchise. The authors point out that with the recent change in intellectual property holder (from Lucasfilm to Disney), Star Wars gained several Brazilian Portuguese translations that for the first time targeted the child audience. The chapter explores the translations of the terms *light* and *dark*, symbolically representing the clash of good and evil, in Star Wars comics, novels, television shows, animations, and games for children. The representation of the ethical dualism in the translation choices adopted is worth investigating in the light of the didactic moral agenda expected from children's cultural products. Translating a transmedia narrative also involves aspects such as narrative continuity, linguistic and technical constraints that become even more challenging owing to the target audience's specific necessities.

Part five explores "intergenerational transmissions," paying special attention to the cross-audience multiplication of voices and visualizations in texts, adaptations, and transmediations that address a variety of different age groups. The contributors analyze how children's literature can speak up in many tongues when read out loud by adults to preliterate young listeners, tackled in crossover fiction, or translated into a "dead language" to provide sophisticated amusement for mature audiences.

Annalisa Sezzi sets out to explore the different voices that can be heard in the first Italian translation and the more recent retranslation of a classic picturebook: *Where the Wild Things are* by Maurice Sendak. Picturebooks are prominently described in terms of the relationship between words and illustrations. Lawrence Sipe defines this interrelation as "synergistic" and the reader's oscillation between the verbal and visual material as a transmediation whereby picturebooks are interpreted. However, their interpretation process is more complex as they come to life in the adults' performance. Thus, the chorus of the discursive presences detectable in children's texts and their translations (E. O'Sullivan, *Comparative Children's Literarure*), such as the voice of the narrator and the translator, is joined by the voice of the adult reading aloud. Since retranslations act as sounding boards for both textual and contextual voices (Alvstad and Assis Rosa 2015), the aim of the analysis is to identify the changes of the voices in the translation (1969) and retranslation (2018) of Sendak's *chef d'œuvre*. The aftermath of the 2018 publication was marked by debates on the new source-oriented translation strategies, culminating in the conference "Celebrating Maurice Sendak" held during the 55th Bologna Children's Book Fair.

Agnes Blümer inspects the difficulties arising from the translation of ambiguity and, more specifically, the translation of dual address in children's fantasy. Her chapter is based on the results of a study dealing with the translation of American, English, and French children's crossover fantasy into German during the decades after the Second World War. Examples include the German translations of Philippa Pearce's *Tom's Midnight Garden*, Mary Norton's *The Borrowers*, Madeleine L'Engle's *A Wrinkle in Time*, Maurice Sendak's *Where the Wild Things Are*, Maurice Druon's *Tistou les pouces verts*, and Eugène Ionesco's *Conte numéro 1*. Blümer conjoins concepts from children's literature studies (children's fantasy, dual address) and translation studies and develops new notions at the intersection of these disciplines: "visual context adaptation" for instances of visual localization in illustrated texts and "generic affiliation" for instances where fantasy texts were adapted to generic models that seemed more established in West Germany at the time. These concepts are used to show how translation was influenced by the ideas surrounding children's literature in that period.

Carl F Miller's title "Maxima Debetur Puero Reverentia": The Histories and Metamorphoses of Latin Translation in Children's Literature might, at first glance, surprise readers with its unexpected juxtaposition of the intellectually elite subject of Latin with the popular field of children's literature. Mapping a rich history spanning over the past 150 years of translating modern children's literature into Latin, Miller realizes a twofold purpose: He illustrates the vitality of a presumably dead language and the "reanimating" potential of the translating activity and challenges the illusory simplicity too often attributed to children's literature. The chapter traces the changing translational intent and target readership of children's literature in Latin translation—from Comenius's 1658 *Orbis Pictus Sensualium,* through the Latinized adventures of Alice, Pinocchio, Struwwelpeter, and Harry Potter, to Alexander Lenard's "paradigmatic" 1960 translation of A. A. Milne's *Winnie the Pooh* as *Winnie Ille Pu*, to the Bolchazy-Carducci Publishers' Latin and Greek publications governed by the aim of "a responsible popularization." Besides scrutinizing significant academic issues, like how trends in Latin education in the English-speaking world have influenced the production of Latin translations of English children's literature, Miller offers examples for cultural hybridity by explaining how and why high-profile translators deal with "lowbrow" stories. *Walter the Farting Dog* transplanted into Latin by a distinguished Classicist or *Diary of a Wimpy Kid* translated by a Vatican cleric aimed to engage the general public with "a familiar text in an unfamiliar language" while "getting kids hooked on the language of Virgil" (Mancini 1). Miller's chapter, too, attests that the dialogue created between dead and living languages can transform educational narratives, and encourage young readers to "resist established ways of thinking promoted through formal schooling" and, hence, provide a way "to sow and nurture the seeds of social change [and] to contribute to developments of equality and diversity" (Reynolds 5).

Caisey Gailey examines three board books that translate scientific knowledge for babies and toddlers. Although children's science books have existed for generations explaining basic concepts and animal-habitat identification among other things, these board books are the start of a recent movement that presents sophisticated concepts such as Newtonian and Quantum Physics. But are these quirky books actually for toddlers or for the parodic amusement of science-minded adults? The chapter argues that—based on considerations of the cognitive requirements of science, use of picturebooks in the acquisition of literacy, analysis of visual and linguistic design elements, the accuracy of information, and the pre-science and future-looking potential of these books—the ostensible purpose lies in encouraging young audiences toward science as they mature.

In today's multimodal environment—where the interaction of words with still and moving images, diagrams, typography, page layout, vocality, music, or corporeal performance are "deployed for promotional, political, expressive and informative purposes"—"technical translators, literary translators, copywriters, subtitlers, localizers, publishers, teachers, and other professionals working with language and text" must learn to account for the relentlessly multiplying signifying elements (C. O'Sullivan 2)[5] to allow child audiences access to their global heritage, and provide a kaleidoscopic reading experience.

Notes

1. Hybrid new subjectivities elicited by new media's interactive potentials have been recently referred to as "prosumers' (Toffler 1980; Ritzer and Jurgenson 2010) or "produsers" (Marshall 2004): compound words made up of the fusion of "producer" and "consumer" to denote the activity

of browsing through media contents to make sensible choices that can eventually prove to be transformative of the meanings to be generated.

2. In a letter to Helen Fielden, written 12 April 1881 Carroll ponders: "*And now what can I say on my own account? Shall I send you a Dutch version of* 'Alice' *with about eight of the pictures done large in colours! It would do well to show to little children. I think of trying a coloured* 'Alice' *myself, a 'nursery edition.' What do you think of it?*"
3. Its twin text is *Cats of Copenhagen*, another fable letter Joyce wrote to his grandson just a few months after he sent him *The Cat and the Devil.* It was published posthumously in 2012 by Scribner illustrated by Casey Sorrow. *The Cat and the Devil*'s illustrators include Richard Erdoes (in the first US edition in 1964), Gerald Rose (in the first UK edition in 1965), and Roger Blachon (in the 1981 Shocken edition, and the 1985 French retranslation by Solange and Stephen Joyce).
4. A perfect illustration of the inseparability of the digital and analogue realms: one can pay a haunting visit to Anne's Secret Annex online in 3D at www.annefrank.org today, by means of a transmedia extension of the diary's reading experience.
5. Carroll O'Sullivan and Caterina Jeffcote co-edited a special issue (July 2013) of *JoSTrans. The Journal of Specialized Translation* on *Translating Multimodalities*. Although the issue had a more general focus, two articles dealt with children's culture in translation by scrutinizing "Translating board games: multimodality and play" and "The interpretation and visual attention of hearing impaired children when watching a subtitled cartoon." http://www.jostrans.org/issue20/issue20_toc.php.

Work Cited

Bassnett, Susan. "The Translation Turn in Cultural Studies." *Constructing Cultures: Essays on Literary Translation*, edited by Susan Bassnett and André Lefevere, Multilingual Matters, 1998, pp. 123–140.

———. *Translation Studies*. Routledge, 2002.

Beeler, Karin, and Stan Beeler. *Children's Film in the Digital Age: Essays on Audience, Adaptation and Consumer Culture*. McFarland, 2014.

Benjamin, Walter. "The Task of the Translator." *The Translation Studies Reader*. Translated by Harry Zohn and edited by Lawrence Venuti, Routledge, 2005, pp. 15–23.

Bishop, Marion. "Confessional Realities. Body-Writing and the Diary of Anne Frank" *Confessional Politics*, edited by Irene Gammel. Southern Illinois University Press, 1999, pp. 13–28.

Bolter, David J. and Richard Grusin. *Remediation: Understanding New Media*. MIT Press, 1999.

Clinton, Katie, Henry Jenkins, and Jenna McWilliams. "New Literacies in an Age of Participatory Culture." *Reading in a Participatory Culture: Remixing Moby-Dick in the English Classroom*, edited by Henry Jenkins and Wyn Kelley. Columbia University, Teachers College, 2013, pp. 3–25.

Cordingley, Anthony and Céline Frigau Manning. *Collaborative Translation from the Renaissance to the Digital Age*. Bloomsbury, 2017.

Cronin, Michael. *Translation in the Digital Age*. Routledge, 2013.

Fetterley, Judith. *The Resisting Reader: A Feminist Approach to American Fiction*. Indiana University Press, 1978.

Geerts, Sylvie and Sara Van den Bossche, editors. *Never-ending Stories: Adaptation, Canonisation and Ideology in Children's Literature*. Academia Press, 2014.

Hennard Dutheil de la Rochere, Martine, Gillian Lathey, and Monika Wozniak, editors. *Cinderella Across Cultures*. Wayne State University Press, 2016.

Hutcheon, Linda and Siobhan O'Flynn. *A Theory of Adaptation*. Routledge, 2013.

Jenkins, Henry. *Convergence Culture*. MIT Press,

———. "Transmedia Storytelling." *Confessions of an Aca-Fan: The Official Weblog of Henry Jenkins*. 2007. http://henryjenkins.org/2007/03/transm edia_storytelling_101.html.

Jenkins, Henry and David Thorburn, editors. *Rethinking Media Change: The Aesthetics of Transition*. MIT Press, 2003.

Joosen, Vanessa and Gillian Lathey, editors. *Grimms' tales Around the Globe: The Dynamics of their International Reception*. Wayne State University Press, 2014.

Kérchy, Anna. *Alice in Transmedia Wonderland*. MacFarland, 2016.

Kirby, Alan. *Digimodernism: How New Technologies Dismantle the Postmodern and Reconfigure Our Culture*. Continuum, 2009.

Kokkola, Lydia. "Interpictorial Allusion and the Politics of 'Looking Like' in Allison and Emma's Rug by Allen Say." *Children's Literature Association Quarterly*, vol. 36, no. 1, 2011, pp. 47–63.

Lathey, Gillian. *The Role of Translators in Children's Literature: Invisible Storytellers*. Routledge, 2010.

Lefebvre, Benjamin, editor. *Textual Transformations in Children's Literature: Adaptations, Translations, Reconsiderations*. Routledge, 2013.

Leitch, Thomas, editor. *The Oxford Handbook of Adaptation Studies*. Oxford University Press, 2017.

Lindseth, John and Allan Tannenbaum, editors. *Alice in a World of Wonderlands: The Translations of Lewis Carroll's Masterpiece*. Oak Knoll Press, 2015.

Mancini, Mark. "10 Popular Children's Books That Have Been Translated Into Latin." *Mentalfloss*, 11 January 2016.

Manovich, Lew. "Understanding Hybrid Media." *Manovich.com*. 2007.

Marshall, David P. *New Media Cultures*. London: Hodder Arnold, 2004.
Mitchell, W. J. T. *Picture Theory: Essays on Verbal and Visual Representation*. University of Chicago Press, 1994.
Moruzi, Kristine, Michelle J Smith, and Elizabeth Bullen, editors. *Affect, Emotion, and Children's Literature: Representation and Socialisation in Texts for Children and Young Adults*. Routledge, 2017.
Müller, Anja. Müller, editor. *Adapting Canonical Texts in Children's Literature*. Bloomsbury, 2013.
Nikolajeva, Maria and Carole Scott. *How Picturebooks Work*. Routledge, 2001.
Nodelman, Perry. *Words About Pictures: The Narrative Art of Children's Picture Books*. University of Georgia Press, 1988.
O'Sullivan, Carol. "Introduction: Multimodality as challenge and resource for translation." *Translating Multimodalities, special issue of The Journal of Specialised Translation*, edited by Carol O'Sullivan and Caterina Jeffcote, no. 20, July 2013, pp. 2–14. http://www.jostrans.org/issue20/issue20_toc.php.
O'Sullivan, Emer. *Kinderliterarische Komparatistik*. Winter Verlag, 2001.
———. *Comparative Children's Literarure*. Routledge, 2005.
Oittinen, Riitta. *Translating for Children*. Garland, 2000.
Reynolds, Kimberley. *Children's Literature: A Very Short Introduction*. Oxford University Press, 2011.
Schulte, Rainer. "Translation and Reading." *UT Dallas. Centre for Translation Studies Website*. https://translation.utdallas.edu/essays/reading_essay1.html.
Serafini, Frank, Dani Kachorsky, and Maria Goff. "Representing Reading: An Analysis of Professional Development Book Covers." *Journal of Language and Literacy Education*, vol. 11, no. 2, 2015, pp. 94–115.
Smith, Howard A. *Teaching Adolescents: Educational Psychology as a Science of Signs*. University of Toronto Press, 2007.
Sipe, Lawrence R. "How Picture Book Work: A Semiotically Framed Theory of Text-Picture Relationships." *Children's Literature in Education*, vol. 29, no. 2, 1998, pp. 97–108.
Stephens, John and Robyn McCallum. *Retelling Stories, Framing Culture: Traditional Story and Metanarratives in Children's Literature*. New York: Garland, 1998.
Sundmark, Björn and Anna Kérchy, editors. Translation and Transmediation Special Issue. *Bookbird: International Journal of Children's Literature*, vol. 56, no. 1, 2018.
Van Coillie, Jan and Walter P. Verschueren, editors. *Children's Literature in Translation: Challenges and Strategies*. St. Jerome Publishing, 2006.
Venuti, Lawrence. "Adaptation, Translation, Critique." *Journal of Visual Culture*, vol. 6, no. 1, 2007, pp. 25–43.
———. *The Translator's Invisibility*. Routledge, 1995.

Waller, Alison. "Translations and Transmediations." *Re-Reading Childhood Books: A Poetics*. Bloomsbury, 2019, 169–175.

Westera, Wim. *The Digital Turn: How the Internet Transforms our Existence*. Author House, 2012.

Wu, Yan, Kerry Mallan, and Roderick McGillis, editors. *(Re)imagining the World: Children's Literature's Response to Changing Times*. Springer, 2013.

Inter-/Intra-Cultural Transformations

Translated into British: European Children's Literature, (in)Difference and *Écart* in the Age of Brexit

Clémentine Beauvais

Three weeks before what has come to be known as the "Brexit" vote, Julia Eccleshare, possibly the most influential children's book reviewer in the United Kingdom, wrote for the Guardian an article on "the best children's books to help children feel connected to Europe." The United Kingdom aside, none of them was actually from a European country, and none was in translation. As if seized by an afterthought, Eccleshare added, "Of course, another option is to read European books in translation. From Tintin to Asterix…".

That such an article, so close to the Brexit referendum, should have been published, let alone written, testifies to the British obliviousness to the fact that it is simply *not normal* for a country in Europe—indeed, for any country in the world—to have so few imports of children's literature from elsewhere. While exact figures are unknown, it is estimated that less than 4% of the children's literature market in the United Kingdom is made up of translations. In many other European countries, that share is well over half. So dire is the state of translated children's books on the island that it does not occur to the most benignly liberal, Europhile,

C. Beauvais (✉)
University of York, York, UK

A. Kérchy and B. Sundmark (eds.), *Translating and Transmediating Children's Literature*, Critical Approaches to Children's Literature,
https://doi.org/10.1007/978-3-030-52527-9_2

cosmopolitan newspaper editors and journalists that such books may be, just possibly, a better way of "helping children feel connected to Europe" than even the best-intentioned homegrown fiction.

Furthermore, Eccleshare's list is holiday-focused and decidedly urban. The city trip is especially foregrounded: "The quickest and simplest way of getting a picture of life in Europe," she argues, "is to look at some of the books about holidays in European capitals," which give "delicious insights into European cities." Four out of nine books take place in Paris—a fact which, while flattering to this Parisian exile, calls into question the existence of Europe beyond the Eurostar. This is a list, clearly, of curated dips into not-too-far-away cultures; the implied British child reader of such fiction is placed in the position, always already, of a contented outsider. Europe is its oyster; yet the continent, it is understood, cannot welcome him or her as a full person—always only as a pair of eyes, camera-wielding hands, and a salivating tongue. The continent, Eccleshare's list implies, must forever for British children remain, however "connected" they feel to it, fundamentally exterior, alien, other—*different*.

In this chapter, I want to probe that *difference*—in all its modalities—of the UK children's books market in relation to European children's literature. Because the situation of children's literature in translation in the United Kingdom is so *different* to the rest of Europe, and because the United Kingdom is currently living through a time of unprecedented political tension with "the continent," I argue that, to study children's literature in translation on the island, we cannot adopt the same analytical and theoretical tools as we would for children's books in translation elsewhere. In the United Kingdom, translating children's books, publishing, promoting, and teaching such books, and, in equal measure, studying them, are activities de facto so vividly anomalous, so stamped by commitment, as to require their own conceptual, aesthetic, and ideological frameworks for analysis.

I propose here that the first step toward this framework might be to be critical of the notion that children's literature in translation exposes, or should expose, children to *difference*. The term is often mentioned but rarely defined, in relation to children's books in translation. In the case of the United Kingdom, it is particularly problematic (as I unpack in the first subpart) because translated literature is always marked already by intractable *difference* on several levels. That phenomenon itself (as I detail in the second part) must be resituated within a long British history of narratives of difference toward Europe, notably as regards language learning.

Yet the "solution" is not empty celebrations of sameness or advocacy of "indifference" to translation; not only would this be probably unfeasible, but it is also arguably undesirable. I contend in the third part of this chapter that the situation calls for theorizations of children's literature in translation more receptive to the tensions and volatility of British people's relationships to Europe and to other languages. I propose Francois Jullien's vision of *écart* (the gap or sidestep) as a conceptual opening toward a more flexible analytical framework, more successfully tuned into the committed aesthetics of children's literature in translation in the United Kingdom.

Vive La Différence? The Intrinsic *Difference* of Children's Books in Translation in the United Kingdom

The question of the imbalance of children's literature in translation in comparison to Anglophone literature is well-known to scholars and education professionals. Nicholas Tucker, in 2005, called it out as a "British problem," noting:

> Abroad is not just about politics; it is also about *different* ways of seeing, feeling and behaving. Continental illustrators ... carry with them an exciting whiff of subversion for readers used only to how things are at home. Authors ... who are translated, do the same thing in print. *Vive la différence!* indeed, but how typical it is that this resounding phrase still as yet has no British equivalent! (emphasis added)

Tucker's cry of "vive la différence" has an ironic twist, as it is in fact a British idiom mimicking a French phrase; indeed, it has no British equivalent—but no French equivalent either. Even in English, the phrase is most often used sarcastically. And while *Vive la différence* may be somewhat of a linguistic mirage, far more puzzling still is its conceptual content when we try to take the expression seriously.

We quickly get into complex theoretical questions when thinking about the question of *difference* in children's literature, and particularly with children's literature in translation. It does seem to go without saying that exposure to cultural or national "differences" is broadly speaking a good thing for children, and that foreign children's literature is particularly able to do that because it is, well, foreign. But there is a faint impression, always, that this exposure to difference is only a first step—that, ultimately, we should hope for that difference to lead to a common understanding. To take another oft-heard metaphor, we intuitively guess that bridges will be built through heightened awareness of the cultural chasm. The rather idyllic vision of translated children's literature as the key to world peace, or at least to some kind of universal understanding, is reminiscent of Paul Hazard's view, in his famous 1932 manifesto, that societies find some common ground through translated children's literature. In Hazard's understanding, child readers are an active community, energized by world literature and shaping it as much as it shapes them, and this movement indirectly benefits humanity, made one by their shared corpus: "Each country gives and each country receives;... and that is how, at the age of first impressions, the universal republic of childhood is born" (231).

However, it is clear that not each country gives its literature, and not each country receives that of others, in even remotely balanced amounts. It is debatable, too, whether children are always the active, discerning corpus-gatherers that Hazard envisages. And even if that were the case, would Hazard's "universal republic of childhood" actually be tolerant, let alone aware, of *difference*? Not really, in Hazard's view at least; it is, if anything, because children's literature is closer to ancestral forms of storytelling that it has universal value. It is worth remembering here the rigidity of the French understanding of "republic" Hazard is calling upon—precisely one that has been seeking, since its inception, to *erase* differences through the process of education. Hazard's view, foundational for the field, was not straightforwardly that translated children's literature was building bridges between people out of their *differences*. He rather envisaged it doing so by bringing everyone closer to a kind of originary similarity. His work set a trend for a more general emphasis on translated children's literature's ability to reconcile, emphasize cross-cultural similarities, and showcase the fluidity of international exchanges. Certainly, the discourse of "translated children's literature as beneficial exposure to difference" has always been a subtext of such academic work in the field; yet it was not always clear by what kind of alchemical operation exposure to difference should transform into a sense of universal belonging.

Furthermore, and even more problematically, even if that kind of magic were true, is translated children's literature truly a guarantee to get the "different ways of seeing" Tucker promises? Can we say, for instance, that children in France are exposed to vast amounts of "difference" because more than half of the production is in translation? How much translation *should* there be in a country, anyway? Let us look at the list of the past ten years' awards of the Prix Sorcières, arguably the most prestigious children's literature award in France. For the mid-grade category, eight out of ten winners are translations; this sounds good, we might venture to say. However, those eight books are translated from *only* three languages. We may begin to worry: is this enough difference? But the authors and illustrators, we note, are of five different nationalities. That does sound like enough difference: it *is* about half. Is half enough difference? Meanwhile, the publishers are not very different: they are all French but for one Belgian. Where is the Francophonie in the Prix Sorcières? But hold on, is this difference only about nationality? Does Katherine Rundell's *Rooftoppers*, while a translation, count as different, since it is set in Paris? And what else should be computed in our *Vive la différence* calculation? How about genders, ethnicities, or social classes represented? Such thought experiments, while worthy in a context marked by increasing concern for voice and representation in children's literature, remain rather futile when the central concepts—difference, diversity—are left untheorized.

Moreover, the Prix Sorcières might reflect the general market statistically, but proportional in number of course does not mean representative in content. In France, translations in the bestselling lists are sensibly the same as those in bestselling lists in the United Kingdom and the United States; they are big UK or US sagas, classics, and books by celebrities. Will this exposure to translation also expose young readers to *difference*, whatever that means? Let us do a quick phenomenology of the child reader in France, exposed from babyhood to bestselling books. Very many in that category will be in translation, which in France is mostly shorthand for "translation from English." By the end of her adolescence, our young reader's attitude to books in translation is very likely to be, if anything, *indifferent*. There will be little in her script as a reader, in her way of apprehending literature, that sets aside the idea or experience of the translated book from the idea or experience of the "normal" book.

Not so in Britain, where an interesting characteristic of the market for books in translation is precisely its *difference* from the market for "normal," namely Anglophone, literature. What is remarkable about the UK market of children's translation is, well, its remarkability. Within the

UK book market, books in translation stand out as abnormal or atypical, and their production is eminently unpredictable and unsystematic. As Gillian Lathey notes,

> Fluctuations in the numbers and sources of translations for children in the UK since 1945 defy any conclusive analysis: economic pressures, chance encounters and the determination of pioneering individuals all play a part in a diverse set of attitudes and practices towards translations in British publishing houses. (232)

Lathey points out the (rather oxymoronic) characteristic serendipity of children's translation in the United Kingdom. "Erratic and sometimes surprising" (232), characterized by "shifting fortunes" (233) and "circuitous [routes]" (235), and reliant on the "enterprising commitment" (234) of individuals, translation for children in the United Kingdom, says Lathey, has providence, coincidence, and chance as its pattern. Namely, it has no pattern.

In an editorial system that is one of the most streamlined and well-oiled in the world, children's translations in the United Kingdom are always already set apart. For them, difference is not an added characteristic: it is their *identity*. If anything, while for the French publishers mentioned above translation is more or less business as usual, most UK publishers who publish translated books can quite rightly stake a claim to *Vive la différence* already.

"*Difference* Thinking" in the United Kingdom's Relationship to Europe

In the fraught relationship between the United Kingdom and Europe, the *difference*-loaded status of children's literature in translation is part of a wider phenomenon that has arguably become more *vive*—alive and lively—since the Brexit vote. At the time of writing, only a year and a half had passed since Brexit; however, current research in neighboring fields to ours, especially language learning, is already looking at the possible reasons for, and effects of, the event.

In a discourse analysis of attitudes to language learning in the United Kingdom since Brexit, Lanvers and Doughy argue that the country has long been marked intractably by—that word again—"*difference* thinking" in relation to language learning. This difference thinking is activated on both personal and collective levels by entangled narratives: "I'm bad at languages anyway" (narrative of individual difference); "as a country, we're bad at languages anyway" (narrative of cultural difference). This is compounded by the perennial problem of English as a dominant language: "the rest of the world speaks English, so we don't need to learn languages anyway" (what might be called a narrative of linguistic difference).

Importantly, Lanvers elsewhere points out that this difference thinking goes both ways. The self-identification of British learners of foreign languages as "bad learners" is compounded by a general perception from Europe and the world that, indeed, the British *are* so ("Contradictory *Others*"). Thus, narratives of difference dominate Britain's thinking about Europe and vice versa. Brexit, Lanvers and Doughy argue, may accelerate this difference thinking by further strengthening the idea that other languages (specifically European) are no longer needed in the United Kingdom; interestingly, they suggest that Brexit also has the potential to reverse the cycle from vicious to virtuous, if the sense of a loss of European identity heightens the urgency of language learning.

Such research also pinpoints the volatility that characterizes the current UK language-learning landscape and that has characterized it for many decades. The fruit of a precarious mixture of policy decisions, cultural Zeitgeist, individual self-perceptions, and geopolitics, the evolutions of language learning in the United Kingdom are fundamentally unpredictable. Here, too, the initiatives of individuals and associations to uphold and celebrate linguistic diversity contrast with successive government decisions—which have been at best sluggish, at worst quite simply toxic to its purposes (for instance, with Labour's decision in 2004 to make language learning optional after the age of fourteen).

In language learning as in children's book translation, there are thus very similar, and seemingly insuperable, narratives of difference. Predictions are difficult; individual initiative is essential; the key to activating the circle virtuously has not yet been found. What Brexit will do to children's literature in translation, as to language learning, remains to be seen. Conversely, though, and for what interests me here, the state of language learning and children's literature in translation gives us some clues to understanding the advent of Brexit. We can, and arguably must, take seriously the hunch that there may be a connection between the cultural, individual, and linguistic "*difference* thinking" in the United Kingdom and the decision made in 2016 by my compatriots.

In this context of conflating narratives of difference and of fundamental unpredictability, children's literature in translation has become, more than ever, a political phenomenon, and the act of translating and distributing children's literature in translation a committed act. Commitment, of course, can be seen as apolitical—editors might *commit* to bringing in different genres, aesthetic tastes, styles, etc. However, considering both the long history of the United Kingdom and its very recent history, what Lathey calls the "enterprising commitment" of individuals to bring children's books in translation to the country has suddenly become more urgent than before and more in need of theorization.

I argue that the situation calls for a change in focus from celebrations of "difference"—there is more than enough difference thinking already at work there—to theorizations of children's literature in translation more receptive to the existing tensions, instability, and unpredictability of the United Kingdom's relationship to Europe and its languages.

Vive L'Écart? Looking into the Gaps

Because of the very peculiar status of children's literature in translation in the United Kingdom, any theorization or claims about its aesthetics must stay attuned to the economic, material, cultural, political, social, and so forth, aspects of its creation and distribution. And because those, as we have seen, are fundamentally unstable, I advocate a flexible conceptual framework: one that does not seek to be totalizing—that does not aspire to systematicity. I want to adjust onto European children's literature in translation in the United Kingdom a theoretical lens receptive to the aesthetics of its multiple commitments.

This means being sensitive to the ways in which text, paratext, epitext, and the conditions of production of children's literature in translation in the United Kingdom exploit, explore, and most importantly perhaps, question and elasticize the difference of those texts from others. Those ways, to reiterate, are not systematic, but mostly erratic; not fixed, but in movement; not the fruit solely of individual intentionalities nor of institutions, but distributed and diffused. That commitment has an aesthetic effect insofar as, on a basic level, it modulates the reception of those texts by readers and conditions their impressions toward those texts in particular. The relationship is dynamic here: I am interested in how this commitment becomes textualized and in how the texts, in return, commit their producers and mediators.

That commitment, prominently, reaches further than single texts, translators, or publishers. We are missing an essential dimension of the specificity of the children's literature "translated into British" if we forget that, precisely because of the irreducible *difference* of translated books in Britain, each example stands not just for itself but for children's literature in translation in general; not just for itself but for the country it comes from; not just for its own language but for the very concept of "another language." The children's book translated into British is characterized by a surfeit of representativeness. That is why we have much to gain by exploring the texts as aesthetically stamped by commitment: they always already signify outwards of themselves to a wider category of text and cultural and linguistic exchanges that are currently under strong political and social debate. To that analytical end, we need tools adjusted both to the enormous variety and fluidity of the corpus and to the instability of all the pivot concepts—namely, the big words of "culture," "representativeness," "difference," "commitment," or indeed "diversity," another often encountered term. I am currently interested in Francois Jullien's concept of *écart*.

Francois Jullien is a French sinologist whose extensive work on Western and Chinese philosophies has given rise to an ample and in some ways controversial work. Central to Jullien's thought is the refusal of traditional comparative work between cultures. Jullien's approach acknowledges the existence of different cultures, with their histories and geographies, but negates the existence of cultural *identity*. He is particularly critical of any endeavor to pin down "cultural differences," which, he argues, fences thought into one or another system. However, Jullien does not suggest that we seek only for similarities—at least not for the benign purpose of cultural conciliation or reconciliation—nor does he advocate searching for universalist set of human properties that would surpass cultures, hovering over them like a divine command.

Rather, Jullien proposes a work of thought happening in the *écart* (the gap, the yawn, the opening) between cultures—not across but in between cultural productions, languages, and perspectives (see, for instance, *Traité de l'efficacité*, *De l'universel*, or "L'écart et l'entre"). He advocates the detour, the "sidestep," by another culture in order to better see what, in one's culture, remains unthought or unthinkable. Shirking from considerations of identity, Jullien's focus invites us to consider the malleable, elastic open spaces between two poles considered as irreducibly different; and to produce meaning from that gap. As he puts it, "While difference is a classificatory concept—difference is the master key of nomenclatures and typologies—*écart* is an exploratory concept, with a heuristic function" ("L'écart et l'entre" 8).

Heuristic is a keyword here. To figure out such questions, we are hindered by clunky concepts, standing like heavy columns to support temples of theory. To navigate that space, we must work from the inside out, through approximations and guesses, hunches and partial attempts leading sometimes to failure and sometimes to illumination. Elsewhere, Jullien posits, difference is a tidying-up concept; *écart* a messing-up concept. Moreover, difference implies fixity, while *écart* tolerates movement. It is the space where thought is deployed across a distance that allows for reflexivity. The concept must resist ethnocentrism and, ideally, dominance. By paying attention to the *écart*, we begin to value those spaces where things happen that tug at our consciousness, making the other culture both graspable and resistant—not blocked behind a glass door but as a place to wander and wonder.

Why would this perspective be applicable to children's translation and the particular case of the United Kingdom? Jullien's thought is born of his encounter with Chinese thought, namely one that, as he pinpoints, has been historically seen as the opposite, the "other," of Western thought—certainly nothing like the relation of Britain to Europe. Yet I want to reclaim it in its more general capacity to get us to think the tensions in the aesthetic work of translated children's literature. Namely, I want to consider the more modest *écarts* offered by children's books in translation as literature and as a cultural phenomenon. Those may be characterized as moments where children's literature in translation yields neither to complete extraneity nor to complete transparency and instead lets its readers and participants hover in an aesthetic in-between—when children's literature in translation, in other words, becomes a tentative proposition to think in-between cultures and languages.

If we start thinking of how translation for children in Britain may allow young readers to feel the *écart* between cultures rather than the difference, we can begin to celebrate that "tugging" relationship to the world that true cultural understanding affords rather than the experiences of pure exoticism or irreducible weirdness (which it also does, indeed, allow for).

Translation for children in the United Kingdom is particularly precious for such an analysis because, to put it bluntly, it is its burden to play with and within the *écart*; it is its fate, both literarily and in market terms, to overthink cultural difference—to dramatize it and place it center stage. If we are to take seriously the notion that Britain's current problems with Europe may be connected to a high tolerance of narrative of absolute difference, on both sides, and little tolerance or appreciation of considerations of *écart*, of in-betweenness, then we can look at translations and translators as some of the most prominent transmitters of a sense of *écart* through language and through the work of translation.

We can do this in various ways. First, we can look at how the texts themselves work elastically with dynamics of alienation and identification, namely how they perform those little detours through another language or culture that Jullien talks about. Practices here are those of close reading, translation analysis, and literary analysis—focusing on the moments when translations pull the English language toward another, making it espouse the contours of another, and conversely, on the moments where translations forego any "translationese" (however controversial the concept might be; see Tirkkonen-Condit). One of my hypotheses is that there may be low tolerance for translated texts or "translationese" in Britain in part because there is low tolerance for a linguistic zone which cannot quite decide on which side of the difference it stands—a language that hovers in the *écart*. Translators, in this case, are key negotiators in that intratextual space.

As way of example for what such an analysis can look like, one could pay particular attention to the ways in which translated children's books in Britain deal with beginnings, that crucial moment for the reader's attachment to the book—readers cracking open the novel at a bookstore, evaluating whether to buy for a child of their acquaintance. The opening of *Toby Alone* (2008), the celebrated translation by Sarah Ardizzone of Timothée de Fombelle's *Tobie Lolness* (2007), is a prime example of a translation's elasticity in that sensitive zone. Ardizzone's very first page works from the source text so freely as to be called more legitimately a version rather than a translation; it offers the reader a far more active Toby than Fombelle's Tobie, swapping most reflexive or passive verbs for active ones, taking out references to Toby's immobility, explicitly assigning thoughts to the hero when they are free-floating in the French text, and indicating straightaway that Toby is being chased:

> Tobie mesurait un millimètre et demi, ce qui n'était pas grand pour son âge. Seul le bout de ses pieds dépassait du trou d'écorce. Il ne bougeait pas. La nuit l'avait recouvert comme un seau d'eau. (de Fombelle)

> Tobie was one and a half millimetres tall, which wasn't big for his age. Only the tips of his feet were sticking out of the hole in the bark. He wasn't moving. The night had capped him like a bucket of water (/water bucket). (Literal translation)

> Toby was just one and a half millimetres tall, not exactly big for a boy of his age. Only his toes were sticking out of the hole in the bark where he was hiding. (Ardizzone)

Later, Toby tells himself, in Ardizzone's version, that the sky in Heaven "couldn't possibly be as deep or as magical as this," activating a readerly script quite different from Fombelle's version, in which Tobie thinks that the sky, in paradise, would be "*moins profond, moins émouvant, oui, moins émouvant…*" (less moving). Whether editorial, translatorial, or more likely a mixture of both, these choices for the opening of Toby's story anchor it quite clearly within a familiar strand of British children's literature—the high or portal fantasy (of the Lewis, Pullman, or Rowling kind). But that anchoring is not nostalgic; in many ways, Ardizzone's beginning is also more resolutely modern than Fombelle's, aligning with a contemporary appetite for in-media-res incipits in children's literature. While Fombelle's Tobie's immobile, quasi-philosophical musing about stars and sky has an Exupéry tinge to it, Ardizzone's Toby, while no less observant of the beauties of the universe, is clearly a little boy on the run, whose body is as present to the text as his contemplation of the sky.

One reading of such a strategy—through the lens of, for instance, Lawrence Venuti's controversial theorization—could be to note, to deplore it or otherwise, the domestication of the source text. Yet this would be unfairly reductive. The translation's contours are not fixed, but shifting, dynamic. Further along, Ardizzone's translation snaps back swiftly to espousing closer the silhouette of Fombelle's text, with the occasional deft, little sidesteps. This translated opening has something to say about the very category of the children's book opening. *Toby Alone* does not simplify, but plays with, stretches, questions, the genre expectations that the French text sets, highlighting with particular vigor by contrast the existing strangeness of that text. Arguably, by overemphasizing in small touches the more conventional nature of Toby's adventure (its genericity, its action-packedness), Ardizzone's text draws special attention also, by contrast, to the delicately alien nature of its aesthetic, to its spiritual considerations, and to its contemplativeness. By doing small detours through action, activeness, and adventure, Ardizzone activates scripts that are joyfully at odds with the rest of the opening; she allows the reader the distance necessary to appreciate, all the more acutely, the dips into thoughtfulness that this opening affords. Therefore, the translation yields neither to foreignization nor to domestication proper; rather, it works dynamically within that space—shifting strategy from one sentence to the next, from one moment of the story (the incipit) to the other. Those translatorial and editorial choices, whether intuitive or strategic, are not reifiable as either packed with difference or comfortably homely. Rather, they work in-between—participating in the complex elaboration, for an audience fundamentally unused to books in translation, of a kind of surprised delight which hints at an *écart* without falling into considerations of *difference*.

This is one kind of analysis only. We can also look at the corpus through distant reading, exploring the publishing system in whatever amount of resistance and giving-in it displays toward the European market. Interesting work by Sinéad Cussen suggests that there is no systematicity in how much publishers play on, and how much they refuse, the label of translation for purposes of promotion. The marshalling of translation within its own category (as is the case for the Marsh Award) or its inclusion within prestigious awards (such as the Carnegie) is an important dimension to consider.

Furthermore, we can look at the friendliness of the British educational and para-educational system to the notion that a book given to children could and should allow for experiences of *écart*. The promotion in the United Kingdom of a "reading for pleasure" agenda (for an overview, see Clark and Rumbold), while doubtlessly precious in many ways, might have implications for translations. What is meant conceptually by pleasure, and the extent to which that definition of pleasure stretches to the potentially destabilizing experiences of reading in translation, must be critically explored. Another prominent and fascinating development of the promotion of translation in the United Kingdom, in the form of translation workshops such as the ones provided by Translators in Schools, is worthy of investigation.

Such approaches amount to reflecting on the acceptance, or lack thereof, of aesthetics of *écart* in the translation of children's literature in the United Kingdom, always bearing in mind the commitment inherent to that type of literature in this country. They require being a committed critic oneself—as well as, if possible, a committed translator, leader of translation workshops, and promoter of translated literature. There is a clear crossover between the work of the scholar and that of the translator in that endeavor. Both have the ability to stand in between two languages and two cultures—placed among the best people to feel that *grand écart* (splits), they are perhaps among the best people to spot it, reflect upon it, and help others wander within it and wonder about it.

References

Ardizzone, Sarah, translator. *Toby Alone*. By Timothée de Fombelle. Walker, 2008.

Clark, Christina and Kate Rumbold. "Reading for Pleasure: A Research Overview." *National Literacy Trust*, 2006.

Cussen, Sinéad. "The Promotion and Reception of Translated Children's Literature in the UK: A Case Study of the Marsh Award." Translation Studies and Children's Literature—Current Topics and Future Perspectives Conference, October 2017, Belgium. Presentation.

de Fombelle, Timothée. *Tobie Lolness*. Paris: Gallimard, 2007.

Eccleshare, Julia. "What are the Best Books to Help Children Feel Connected to Europe?" *The Guardian*, 6 June 2016, https://www.theguardian.com/childrens-books-site/2016/jun/06/best-childrens-books-on-europe-katherine-rundell-ludwig-belman.

Hazard, Paul. *Les Livres, les enfants et les hommes*. Paris: Flammarion, 1932.

Jullien, François. *Traité de l'efficacité*. Paris: Grasset, 1997.

———. *De l'universel, de l'uniforme, du commun et du dialogue entre les cultures*. Paris: Fayard, 2008.

———. "L'écart et l'entre. Leçon inaugurale de la Chaire sur l'altérité." *Pratiques*, 253, 2012.

Lanvers, Ursula. "Contradictory *Others* and the *Habitus* of Languages: Surveying the L2 Motivation Landscape in the United Kingdom." *The Modern Language Journal*, vol. 101, no. 3, 2017, pp. 517–532.

Lanvers, Ursula and Hannah Doughy. "Brexit and the Future of Language Learning in the UK: A Critical Discourse Analysis of Public Discourses." Conference presentation, 2016. Unpublished.

Lathey, Gillian. "Serendipity, Independent Publishing and Translation Flow: Recent Translations for Children in the UK." *The Edinburgh Companion to Children's Literature*, edited by Clémentine Beauvais and Maria Nikolajeva. Edinburgh: Edinburgh University Press, 2017, pp. 232–244.

Tabbert, Reinbert. "Approaches to the Translation of Children's Literature: A Review of Critical Studies Since 1960." *Target*, vol. 14, no. 2, 2002, pp. 303–351.

Tirkkonen-Condit, Sonja. "Translationese—A Myth or an Empirical Fact? A Study into the Linguistic Identifiability of Translated Language." *Target. International Journal of Translation Studies*, vol. 14, no. 2, 2002, pp. 207–220.

Translators in Schools, http://translatorsinschools.org/.

Tucker, Nicholas. "Why is there a British problem?" *Outside In: Children's Books in Translation*. Milet Publishing, 2005. Available at http://www.outsideinworld.org.uk/childrens-books.asp?page=publications-outsidein.

Venuti, Lawrence. *The Translator's Invisibility: A History of Translation*. Routledge, 1995.

———. *The Scandal of Translation: Towards an Ethics of Difference*. Routledge, 1998.

Picturebooks in a Minority Language Setting: Intra-Cultural Transformations

Hannah Felce

Writing, translation, and publishing practices often follow more complex patterns than the "full transposition of *one* (monolingual) source code into *another* (monolingual) target code" (Meylaerts 5). This is particularly true in the case of translators and writers who originate from countries or regions where two or more languages interact within one cultural system. In such areas, the publishing process itself may not be monolingual—since multilingual authors move fluidly between multiple tongues, frequently disregarding boundaries between different languages. Publishing houses may themselves work in multiple languages. In the case of minority language publishers, a work may also be published in multiple forms for different reasons: not just for commercial purposes but also for language maintenance and preservation purposes. Additionally, when looking at children's literature and picturebooks in particular, this process is further complicated by the relationship between the text and illustrations. Moreover, when not one dialectal variety of the minority language dominates—as is the case with Romansh, Switzerland's fourth national language—intralingual textual transformation (conjoining Roman Jakobson's and Benjamin Lefebvre's terms) of a text (i.e., the reinterpretation

H. Felce (✉)
Cardiff University, Cardiff, UK

A. Kérchy and B. Sundmark (eds.), *Translating and Transmediating Children's Literature*, Critical Approaches to Children's Literature,
https://doi.org/10.1007/978-3-030-52527-9_3

of the text from one dialect into another) adds to the multilingual nature of the publishing process. Thus, assumptions regarding language, linear processes in translation, and so-called originals are destabilized.

In this chapter, one of the key questions I will be asking is whether the publication process of children's literature written in a minority language and the interrelationship between the illustrations and text during this process call into question the notion of a predetermined original, as well as the binary model of the translation process and, ultimately, the fixed nature of language. As a case study, I will analyze a picturebook—Selina Chönz's *Uorsin* (1945)—that was simultaneously released in two languages (one a major language and the other a minority one) and in multiple dialects (of the minority language). I will also discuss *Uorsin*'s two sequels, *Flurina und das Wildvöglein* (1952) and *Der grosse Schnee* (1955)—which were first written in German as accompaniments to the images of the illustrator, Alois Carigiet, and were only translated into Romansh at a later point in time. Although research has been carried out on *Uorsin* and its sequels, studies often focus on subjects other than translation, such as Carigiet's illustrations (Hans ten Doornkaat) or the depictions of the two main characters, Uorsin and Flurina (Ofelia Schultze-Kraft), disregarding the "bilingual" nature of the Romansh versions. This means that an analysis of the language ideology behind the intralingual translations has so far been neglected. I will draw on scholarship in translation studies (Roman Jakobson; André Lefevere), adaptation studies (Linda Hutcheon), children's literature (Maria Nikolajeva and Carole Scott; Benjamin Lefebvre), and children's literature translation studies (Emer O'Sullivan; Riitta Oittinen; Gillian Lathey) in order to show that the various linguistic version of *Uorsin* and its illustrations are already in conversation with each other during the publishing process.

Uorsin/Ursin/Uorsign/Uorset/Schellen-Ursli: A Picturebook Released Simultaneously in Five Different Versions

Swiss author Selina Chönz (1910–2000) and Swiss illustrator Alois Carigiet's (1902–1985) *Uorsin* (1945) is one of the biggest lasting successes among Swiss picturebooks (*Neue Zürcher Zeitung* 54). That success was largely due to the dual appeal of the book as, first, an adventure story for children and, second, an atmospheric tableau of Alpine life for adults (Berg-Ehlers 45). The book has been published in ten languages, with the total number of copies sold worldwide thought to be 1.7 million. In 1948, Chönz and Carigiet received the Schweizer Jugendbuchpreis for *Schellen-Ursli* (1945),[1] and in 1953, the English translation won the New York Times Choice of Best Illustrated Children's Book of the Year (Schultze-Kraft 161). The original publisher of the German edition was the *Schweizer Spiegel Verlag* and the *Lia Rumantscha* was and remains the publisher of the Romansh editions. In 1971, all rights were transferred to *Orell Füssli Verlag* (OF). In 1992, OF reviewed the children's book section of their catalogue and only kept this trilogy in publication because of its lasting success. Furthermore, Carigiet, who was a well-known graphic designer in Switzerland, was the first winner of the Hans Christian Andersen Award for Illustration in 1966 for his work for *Uorsin* and its sequels as well as his own picturebook trilogy.

Uorsin was originally written in one of the dialects of Switzerland's fourth national language, Romansh, which is only spoken in Canton Graubünden.[2] It was written in Ladin, the dialect spoken in the Upper Engadin, and was illustrated by another Romansh speaker, Alois Carigiet. He, however, came from the Surselva area and spoke the Romansh dialect known as Sursilvan. Due to Graubünden's mountainous geography and scattered population, as well as cantonal and local autonomy, different languages or dialectal varieties are spoken even between neighboring villages. Romansh is divided into five regional varieties, which Romansh speakers call *idioms*,[3] because each variety possesses its own written version with its own grammar and lexicon. The five idioms of Romansh are Sursilvan, spoken in the Surselva along the Vorderrhein; Sutsilvan, in the Hinterrhein valley; Surmiran, in central Graubünden; Putèr, in the Upper Engadin; and Vallader, in the Lower Engadin and Val Müstair (Cathomas 89). Putèr and Vallader fall under the umbrella term Ladin. The differences between the idioms on a lexical, grammatical, and intonational level are considerable, and speakers of one idiom do not automatically understand speakers of the others. This means that intralingual translations of popular books are required between Romansh variants. These are usually completed prepublication, if at all.

The background of *Uorsin*'s publication is closely linked with the political climate at the time. In nineteenth-century Switzerland, liberals desired to do away with Romansh because they saw it as an obstacle for the Canton Graubünden's future in the modern world. A counter-reaction to this was the *Renaschientscha retorumantscha* (Romansh Renaissance), which grew during the first half of the twentieth century. In the years after the Second World War, many Romansh books were published and became an element of identification and propaganda for the Romansh Renaissance. According to Rico Valär, one of the most successful projects that came out of the Romansh Renaissance was precisely *Uorsin* (38). Valär also argues that, due to its content, *Uorsin* was distributed by the members of the linguistic renaissance specifically because of its strong self-awareness as Romansh literature (38). This is clear in *Uorsin*'s storyline, which is about a young boy from the mountains who searches for the largest bell in the village in order to lead the procession at the regional folk festival, called Chalandamarz, during which the winter is expelled by the sounding of cowbells rung by either schoolboys or school children. That *Uorsin* is an excellent example of Romansh literature is also visible in the paratextual elements of the book, since it reveals the key contributors of *Uorsin* and their agency (Fig. 1).[4]

Fig. 1 Cover illustration of *Schellen-Ursli*

Three Books, Four Prefaces, Five Author-Translators

Chönz approached Carigiet several times before he finally agreed to illustrate her children's story in 1939 (Trullmann). Significantly, this happened to be one year after Romansh was declared a national language of Switzerland, and Carigiet began working on the illustrations for *Uorsin* in 1940. Both Chönz and Carigiet were involved in the promotion of Romansh before they collaborated. Even before she wrote *Uorsin* and its sequels, Chönz wrote stories about rural life in the Alps: *La chastlauna* (1940), *Il purtret de l'antenat* (1943), and *La scuvierta da l'orma* (1950). Furthermore, Chönz's husband, Jachen Ulrich Könz, was a well-known advocate for Romansh culture and language from the Engadin. During the 1930s, Carigiet illustrated multiple front pages for the *Schweizer Spiegel* (coincidentally also the sister newspaper of the original publishing house of *Schellen-Ursli*, the German version of *Uorsin*), which championed the *Geistige Landesverteidigung* (intellectual defense of the nation) (Valär 36). In 1933, *Schweizer Spiegel* also published a contribution by Otto Gieré, who called for Romansh to become a national language in the *Bundesverfassung* (Swiss Federal Constitution). The piece was accompanied by an illustration by Carigiet, which exemplifies Carigiet's political standpoint toward the Romansh language and culture. According to Chasper Pult ("Ein Engadiner Kuppler"), it was at the suggestion of Jon Pult, a Romansh author actively engaged with promoting the Romansh language and culture and the author of both the German and Ladin prefaces of *Uorsin*, that Chönz contacted Carigiet to illustrate the story. Having also grown up in a Romansh-speaking area, Carigiet could thus play a mediating role between the culturally and linguistically diverse Romansh valleys (Pult, "Ein Engadiner Kuppler" 40).

Three editions of *Uorsin* were released simultaneously in 1945: one in German, titled *Schellen-Ursli* and self-translated by Chönz, and two in Romansh. Each of the latter was an "ediziun comünaivla" (Chönz, *La naivera/La cufla gronda*), a joint edition[5] containing two of the Romansh idioms. One Romansh edition comprises the story in Ladin and Sursilvan, and another in Sutsilvan and Surmiran. This complex language scenario is clearly stated on the copyright page of the first edition in German (*Schellen-Ursli*, 1945). Therefore, there are four Romansh versions of *Uorsin* and each was written by a different author-translator.

Two well-known Romansh writers and a language activist were asked to produce the intralingual Romansh translations of Chönz's Ladin version: Catholic priest Gion Cadieli (1876–1952) wrote the Sursilvan version; the well-known Capuchin priest Alexander Lozza (1880–1953) wrote the Surmiran version; and Curò Mani (1918–1997) wrote the Sutsilvan version. All three authors of the intralingual translations were important figures in the Romansh Renaissance. Lozza is one of the most well-known Surmiran authors (Müller) and an important figure in the Romansh Renaissance in Surmeir (Deplazes). Gion Cadieli was also an important figure in the Romansh Renaissance in Surselva (Deplazes). Curò Mani trained as a teacher but was also the main Sutsilvan poet of the Romansh Renaissance and an important author in Schams (Krättli). Furthermore, Mani was a Romansh language activist and worked alongside Giuseppe Gangale at the *Acziun Sutselva rumàntscha*, the movement to reintroduce the Sutsilvan idiom. The cultural capital these authors commanded within the valleys in which their idiom is spoken and their activities aimed at encouraging Romansh language maintenance would persuade adults who share similar political views toward the Romansh language and culture to buy the book for their children.

Taking into consideration the backgrounds of Chönz and Carigiet—as well as those of Pult, Lozza, Cadieli, and Mani—it is clear that there is a political reason for the creation of *Uorsin* and a political agenda behind its publication. This is not only reflected in the roles played by each contributor beyond the production of *Uorsin*, it is also reflected in the multilingual nature of the publication of this book: Translating *Uorsin* into all Romansh idioms would bring together the idioms and ensure as wide a Romansh readership as possible. This function is stated in some, if not all, of the prefaces of *Uorsin*.

The linguistic complexity of *Uorsin*'s publication process is also reflected in the paratexts. As mentioned above, there were several contributors who revealed their agency in the prefaces they author. The authorship of four prefaces—two Ladin prefaces, an original one for the joint 1945 Ladin/Sursilvan version and a later one for the 1963 edition; and two German prefaces, an original one for the 1945 edition and a later one for the 1971 edition—is attributed to Jon Pult. In addition, Alexander Lozza and Curò Mani both wrote a preface for the 1945 joint Surmiran/Sutsilvan edition of *Uorsin*. All *Uorsin*'s prefaces were written by someone other than the main text's author, that is, "allographic prefaces" (Genette). There are two main functions of an allographic preface: recommendation and presentation (Genette 268). In the case of an original preface, "this support is generally provided by a writer whose reputation is more firmly established than the author's" (268). In *Uorsin*'s case, this support was provided by Jon Pult—a renowned author, language activist, president of the *Union dals Grischs* (the language maintenance organization for the Ladin idiom), and the secretary of the Lia Rumantscha (the umbrella organization for the Romansh language and publisher of *Uorsin*). Pult saw Romansh literature as intrinsically important for Romansh culture (Hofmann 2011). Chasper Pult also states that the fact that two Romansh speakers—Chönz and Carigiet—collaborated on the project was probably the main reason for Jon Pult's engagement in the publishing of *Uorsin* ("Ein Engadiner Kuppler" 42). The maintenance of the Romansh language and culture was his life's work; thus, his Ladin preface for the 1945 edition is written as a call for keeping the Romansh language alive, giving something to the future Romansh-speaking generations, and instilling some pride in their heritage as Romansh speakers:

Mo quant plü dastrusch toccarà l'istorgia ils cours da noss pitschens Engiadinais chi sun svessa its cun zampuogns e s-chellas tras las giassas dals cumüns. Schi dain ad els quaist bel regal ch'els possan ir insembel cun „Uorsin da la s-chella" sunand e chantand–portand in lur cours la prümavaira rumantscha. (Pult in Chönz, *Uorsin*, 1945)	[But closer to home, the story will touch the hearts of our little Engadin children, who themselves walked down the villages' lanes with cow bells. So, give them this wonderful gift, so that they can go ring and sing together with Uorsin and his bell, carrying the Romansh spring in their hearts.]

However, in order to achieve these goals beyond just the Ladin-speaking population, intralingual interpretations into the other idioms were required. These intralingual textual transformations of *Uorsin* are mentioned in Pult's preface, as is Chönz's German self-translation:

Il cudesch ais stat tradüt cun prontezza eir in oters idioms rumantschs. Sur Gion Cadieli ha fat l'adattaziun sursilvana dasper il text ladin. In ün'ediziun a part cumpara *Uorsin* surmiran da Pader Alexander Lozza ed ün *Uorset* sutsilvan da Curò Mani. A medem temp vain oura ün *Schellen-Ursli* tudais-ch e preparà ün *Ourson* frances. (Pult in Chönz, *Uorsin*, 1945)	[The book was also promptly translated into other Romansh idioms. Mr Gion Cadieli created the Sursilvan adaptation accompanying the Ladin text. In another separate edition, an *Uorsin* in Surmiran by Father Alexander Lozza and an *Uorset* in Sutsilvan by Curò Mani was published. At the same time, a German *Schellen-Ursli* was published and an *Ourson* in French is being prepared.]

According to Genette (264, 268), in the case of a translation, the preface is generally provided by a writer who is better known in the importing country or by the translator. Lozza and Mani fulfill both these roles in the double preface of the 1945 Surmiran/Sutsilvan joint edition; they are both better-known authors in the importing idiom and the translators of the text. In this joint edition, both Lozza and Mani write a preface to *Uorsin*—one in Surmiran, the other in Sutsilvan, respectively. These two prefaces appear under the title "Dus pleds oravant" (Chönz, *Uorsin*, 1945), which reflects Pult's 1945 preface titled "Duos pleds sün via" (Chönz, *Uorsin*, 1945). Lozza's preface reflects Pult's encouragement for language preservation. However, it does not highlight the multilingual nature of the publication of *Uorsin*. Instead, it highlights the fact that it is available in the Surmiran idiom: "Egn tgi è ai sez cun la stgella, da Calonda-mars, porscha ena libra versiung an rumantsch-Surmeir, dallas aventuras d'Ursign dalla stgella" (For those who went around with a bell for Chalandamarz themselves, there's a free version in Surmiran Romansh of the adventures of Uorsin of the bell). It also highlights that the author is from another Romansh-speaking area: "L'ancunaschainta scribenta ladina, Selina Chönz" (The unknown Ladin writer, Selina Chönz). In comparison, Mani's preface specifically mentions the multilingual background of *Uorsin*'s publication and the two contemporaneous editions to the Surmiran/Sutsilvan joint edition:

Igl cudisch e vagnieu stampo an quater idioms rumantschs, an tudestg ad an franzos. Gest nus da la Sutselva vegn nerdabasegns dad el. Egn tgòld angraztgamaint alla Leia Rumàntscha ca e vegnida ancunter a nus. Possi igl cudisch catar tut igls cors da Viulden a Prez veiadaint tras las Veiasmalas a tras Schons tocan Calantgil. (Curò Mani in Chönz, *Uorsin*, 1945)	[The book was published in four Romansh idioms, in German and in French. Even we from the Sutselva need it. A warm thanks to the Lia Rumantscha for making it available to us. May the book find its way into all the hearts from Veulden to Präz, through Viamala and Schams to Innerferrera.]

The way the idioms were divided up between the volumes shows that there are other factors at play besides the attempt to reflect the linguistic diversity of Romansh. The idioms are not placed together in the editions based on the degree of mutual understanding between them, but rather on the authors' backgrounds. Chönz originally wrote *Uorsin* in Ladin, and Sursilvan is the idiom spoken by Carigiet. Thus, publishing a joint edition in these two idioms meant that the Lia Rumantscha, the publisher of the Romansh volumes, could maximize sales for the book within the valleys from which the two contributing authors originated. By default, the remaining two idioms, Sutsilvan and Surmiran, were published together in the other joint Romansh edition. Additionally, by publishing joint editions, the Romansh volumes could be printed in a calculable number of copies (Pult, "Kastanien mit Schlagrahm" 51). The decision to place two idioms alongside one another was also based on economic factors: Four versions of the same story by four writers accompanied by illustrations by one famous graphic designer would make for the widest readership possible in Romansh, and one edition containing two versions in different idioms could be sold in two different areas that spoke different versions of Romansh. That this is largely an economic factor is apparent from the publishing of the third book in the trilogy, *La naivera* (1964), which was the first of Chönz's books to be published in an edition containing one idiom only, Ladin. The only other idiom in which it was published was Sursilvan as *La cufla gronda* (1964).

Furthermore, editions containing two versions in two variants of the language arguably enable Romansh readers to gain access to the other idioms of Romansh, thereby strengthening the relationship between those idioms. Many Romansh books were published in the postwar years as a means of promoting Romansh cultural identity as part of the Romansh linguistic renaissance. These publications became an element of identification and propaganda. However, not all were translated into all the idioms. According to Valär, one of the most successful projects that came out of the Romansh renaissance was precisely *Uorsin*. It was due to its strong self-awareness as Romansh literature and its portrayal of Romansh culture that *Uorsin* was embraced and distributed by members of the Romansh linguistic renaissance (Valär 38). It is also highly probable that the use of well-known authors as translators to author their own interpretations of the text in the other Romansh idioms contributed to its success, as did Carigiet's renown and his well-loved illustrations, since these author-translators could thus advocate for Romansh in their respective valleys. Consequently, their intralingual textual transformations are not translations in the traditional sense but rewritings that reflect on and extend the original in a deliberate manner to preserve the Romansh culture (Hutcheon xiv, 4; Brodzki 1–2). This is reflected in the language used in both the Romansh and German editions. As seen from the examples above, the paratexts of the editions include words such as *Übertragung* (transfer) into German and *adataziun* (adaptation) into Romansh. In this case, however, both the German interlingual and Romansh intralingual transformations occur prior to publishing and, thus, form an integral part of the "original," giving equal status to the three simultaneously published versions.

Nevertheless, the layout of the joint editions clearly separates the two idioms within each couple and maintains Chönz's authorship of the text. The Ladin text is followed by the Sursilvan text in italics, separated only by the drawing of a flower. On the title page, Chönz and Carigiet's names are placed next to each other under the title; in the Ladin/Sursilvan edition, Gion Cadieli's name is written below those of author and illustrator in a smaller font and accompanied by the statement that his work is an adaptation: "adattaziun sursilvana: Sur Gion Cadieli" (Chönz, *Uorsin*, 1945). This layout is also echoed in the Sutsilvan/Surmiran edition. Nevertheless, the fact that these versions were published simultaneously places the adaptations on an equal footing with Chönz's Ladin version. Speakers of Sursilvan, for example, would focus on their idiom instead of Chönz's version—first, because of their greater familiarity with Gion Cadieli over Chönz, since at the time he was more well known in the Surselva area, and second, because Ladin would not necessarily be accessible to them. On the other hand, the fact remains that the joint editions allow a curious child or adult to compare the different dialects of their own language if familiar with different idioms.

A Multilingual or Monolingual *Schellen-Ursli*?

As already mentioned, besides the intralingual textual transformations into other Romansh idioms, a German edition of the text was also published contemporaneously with the source text. This German version later became the one from which further translations into other languages were made (Pult, "Kastanien mit Schlagrahm" 50). In this section, I will discuss the translation of the text into German and whether the multilingual nature of the creation and publishing processes of the Romansh editions is also reflected in it.

As stated in the German edition's preface of *Schellen-Ursli*, the German text is a free translation of the original, written by Chönz herself (Schultze-Kraft 159). Initially, Chönz was not going to create the German version herself, but she was dissatisfied with the versions submitted by the two German authors approached for the translation (Pult, "Kastanien mit Schlagrahm" 54). That *Schellen-Ursli* (1945) is a self-translation by the author is also mentioned in the preface of the German first edition (1945): "Die Autorin hat ihren romanischen Text frei ins Deutsche übertragen" (the author translated her Romansh text freely into German). This gives Chönz's German version equal status to her Ladin version and shows the multilingual competence of the author. Yet it ignores the more complex multilingual background of Romansh and its translation into the other Romansh idioms. However, in the copyright pages of the German edition, the larger multilingual endeavor of *Uorsin*'s publishing is highlighted. Two of the three editions published in 1945—the Ladin/Sursilvan joint edition and the German edition—mention the other linguistic editions published contemporaneously. This explicitly draws the reader's attention to the multilingual processes involved in the production of the Romansh and German versions of *Uorsin*:

Von dem Buch "Schellen-Ursli" erscheinen unter dem Titel "Uorsin" im Verlag der Ligia Romontscha in Chur *gleichzeitig*[6] zwei romanische Ausgaben; eine mit ladinischem und surselvischem und eine zweite mit surmiranischem und sutselvischem Text (*Schellen-Ursli*, 1945).	[From the "Schellen-Ursli" book, two Romansh editions are being published under the title "Uorsin" at the same time by the publishing house of the Lia Rumantscha; one in Ladin and Sursilvan and a second one in Surmiran and Sutsilvan.]
Üna ediziun tudais-cha dal "Uorsin" cumpara suot il titul "Schellen-Ursli" *a medem temp*[7] pro l'editur Schweizer-Spiegel a Turich. / Ina ediziun tudestga digl "Ursin" cumpara sut il tetel "Schellen-Ursli" *a medem temps* tier igl editor Schweizer-Spiegel a Turitg. (*Uorsin*, 1945) (Ladin/Sursilvan version)	[A German edition of "Uorsin" titled "Schellen-Ursli" is being published at the same time by the editor Schweizer-Spiegel in Zurich.]

Yet, the Sutsilvan/Surmiran joint edition does not mention the other editions on its copyright page.

Moreover, other than in the instances above, the books do not reflect the linguistic diversity of Canton Graubünden and of the author. When talking about the multilingual competence of the author, we must not only keep in mind her ability to speak and write in both German and Romansh but also refer to her competence in both High German and Swiss German, or *Mundart*, since in Switzerland forms of diglossia also exist between these two varieties of German.[8] Chönz's self-translation, *Schellen-Ursli*, was written in standard German, and Swiss German is not used in any form within the German edition. A joint edition in German containing a Swiss German and a High German variant could have reflected the diglossic situation in Switzerland. Instead, it was left to the adult reader to translate the written High German into spoken Swiss German for the child if they wished to do so (Studer 91). This is the usual way such an "intralingual translation" is done in Switzerland, and only a very few Swiss German texts or translations into Swiss German from High German have been introduced on the Swiss literary market (Studer 81). However, this does not mean that the linguistic diversity also present in the German-speaking area of Switzerland is not at all present in the book: *Schellen-Ursli* just shifted multilingualism and the translation process elsewhere (i.e., to the role played by the adult reader).

The Relationship Between Illustrations and Text

Literary illustration as translation is a complex process of cross-temporal, cross-spatial recontextualization. Using research in the translation of children's literature (O'Sullivan)—which is in turn based on Jakobson's notion of intersemiotic translation (233), or translation as an interpretation of verbal signs by means of signs belonging to nonverbal sign systems—this section aims to demonstrate how literary illustration is a form of translation. It also aims to show that if translation is viewed as a form of rereading or rewriting (Lefevere; Oittinen)—something that is "a dialogic, carnivalistic, collaborative process carried out in individual situations" (Oittinen, "The Verbal" 84), "where illustrators, authors, translators, publishers, and different readers meet and influence each other" ("Where the Wild" 129)—the publishing process of a picturebook created by an author-illustrator team questions the notion of a predetermined original. What would usually be viewed as the original (i.e., the published book) is the product of negotiations (or influences) that take the form of intersemiotic translation and textual transformations introduced in the negotiation between the author and illustrator. In *Uorsin*'s case, as we will see, the text influenced the illustrations, which then influenced the text in return. The illustrations and text, taken together as a single published entity (iconotext), then influence the reading of the picturebook (Nikolajeva and Scott 2). This is a continuous, never-ending cycle of what Oittinen (*Translating* 138–139; "The Verbal" 96)—drawing on the work of Bakhtin (124–125)—refers to as crowning and uncrowning. In other words, one day the author is "the symbol of authority" or queen/king, the next day, the author loses her/his authority and the illustrator becomes the queen/king.

The Ladin text for *Uorsin* was written by Chönz first, and only once she had a "finished" version did she approach—or crown—Carigiet to illustrate the story with images. After some hesitation, Carigiet agreed to take on the task because Chönz's use of his childhood language moved him to do so: "My work in this field [children's literature] was inspired by the written word, in particular that of the Romansch language. I was first inspired in this direction by words uttered in my first meeting with Selina Chönz" (Carigiet 157). It then took him five years to produce the illustrations. Over that period, he made multiple visits to the village in the Engadin, where Chönz lived. This village, called

Guarda, alongside Chönz's Romansh words, became Carigiet's inspiration for the illustrations. Thus, the illustrations are a form of translation stemming from a specific environment, as well as from Chönz's words, and from the Romansh language in general or, more specifically, what Carigiet perceived the Romansh language to be—that is, the Sursilvan idiom with which he was familiar.

Yet the (re)visualization of *Uorsin* did not end there. Once Carigiet had created the illustrations, Chönz altered certain parts of the text to better match the images. Once again, she is given "the symbols of authority" (Bakhtin 124; Oittinen, *Translating*; "The Verbal" 96). In other words, she transformed or translated her text to better align with the illustrations. This is especially reflected in the length of the story since Chönz shortened it to fit in with Carigiet's illustrations: "Il text original sto gnir scurzni da manü" (the original text had to be substantially shortened) (Trullmann). That Chönz made alterations to the text based on the illustrations is also apparent in the German version of *Uorsin*. In the published Ladin version, the reader finds many of *Uorsin*'s characteristics described in the written text, which is then used by Carigiet as inspiration for his illustrations—something we can see in the illustrations themselves. For example, in the Ladin version Chönz describes Uorsin as smiling, with brown hair, wearing a belt made by his father, and we are told that the color of his hat is fading toward the tip so that everyone recognizes him. However, these descriptions are not present in the German version, where Chönz (who self-translated the story) instead allows the reader to notice these characteristics only through the book's illustrations.

In fact, when analyzing the various Romansh versions, it is clear that the Romansh intralingual translations also used both Chönz's German text and Carigiet's illustrations for inspiration. For example, in the description of Uorsin, the Sursilvan version follows the same structure and provides similar information in the same rhyming couplets as the German text, and the Sutsilvan version makes reference to Uorsin's black hair as in Carigiet's illustration, when it is clearly described as brown in Chönz's Ladin text: "El ria, sguerschagiand adüna /suotour sieu clap chavlüra brüna" (He's smiling, as always with a cheeky sideways glance from under his thick brown hair). That the author-translators were influenced by both Chönz's Romansh and German texts as well as Carigiet's illustrations shows that this process of crowning and uncrowning is fluid and involves several contributors and stages. Both on a macro and micro level, the illustrations and written text of a picturebook are always in conversation. When copresent in a work, they form separate parts of the whole and, thus, perform different functions. This double layer of intersemiotic translation is present in the process and product of the translation of *Uorsin*, as well as its sequels.

The Sequels: *Flurina and the Wild Bird* and *the Snowstorm*: A Different Approach

Although Chönz and Carigiet had not originally planned to publish a trilogy, the two sequels to *Uorsin* were clearly produced with *Schellen-Ursli* in mind, and the move to create a coherent trilogy was very deliberate. Except for the addition of small sketches to the main illustrations, the style remains the same and several motifs are used throughout the trilogy. Following the success of *Uorsin*, Chönz and Carigiet decided to publish a sequel titled *Flurina und das Wildvöglein* (1952), a story about Uorsin's sister, Flurina. Shortly after, *Der grosse Schnee* (1955), a story about the two siblings set in the alpine winter, was also published. Originally, *Uorsin* was written as a stand-alone story, and only after the book's huge success did Chönz and Carigiet decide to create the other two stories. In the case of the two sequels, however, the order in which the different language versions were produced is the opposite to *Uorsin*: Chönz first compiled *Flurina und das Wildvöglein* and *Der grosse Schnee* in German and then translated them into Romansh (Figs. 2 and 3). Additionally, in comparison to the production of *Schellen-Ursli*, Carigiet describes the two sequels as "Bilderbücher mit begleitendem Text" (illustrated books with accompanying text) (Carigiet 63). After a brief discussion with Chönz about the overall plotline of the book, Carigiet created the illustrations based on his interpretation of the story, and only then did Chönz write the text to accompany the illustrations. Thus, there is a reversal here to the usual tendency to see illustrations in picturebooks as translations of the text. Instead, here the text becomes a transcript of the illustrations. In other words, the process of crowning and uncrowning, or influence (Oittinen, *Translating*; "The Verbal"), here unfolds in reverse order in comparison to *Uorsin*. This once again leads us to question the very notion of translation as a binary activity moving from one singular code into another singular code. In fact, in our case, we have an exchange—a retelling—and the translation is especially not singular since the plotline influenced the illustrations and the illustrations influenced the text.

Fig. 2 Cover illustration of *Flurina und das Wildvöglein*

Furthermore, unlike *Uorsin*, in the case of both sequels, the Romansh and German versions were not published simultaneously. In fact, the Romansh versions of *Flurina* (1953) were published a year after the German book (1952), and the Romansh versions of *Der grosse Schnee* (1955) were only released nine years after the German edition in 1964 (Schultze-Kraft 160). The English (1953, 1961), French (1955, 1956), and Japanese (1954) translations of the sequels even preceded the Romansh editions. This delay between the publication of the German version of the sequels and their Romansh translations shows that, in this case, producing translations into the Romansh idioms was not considered as important as having *Uorsin* immediately available in German. Since Chönz was bilingual and thus able to write in German, too, the Romansh text became superfluous to the publisher.[9] This change in language reflects both the power relations between the minority and majority language and the status acquired by the German free translation

Fig. 3 Cover illustration of *Der grosse Schnee*

of the first book. Thus, the translation process is reversed in the case of the sequels since the minority language, Romansh, is completely foregone in the initial publication and is only added *a posteriori*.

Moreover, in the case of the Romansh editions of *Der grosse Schnee* (1955), the Ladin and Sursilvan versions were not published in a joint edition. Instead, they were published individually and titled *La naivera* (Chönz's self-translation) and *La cufla gronda* (adaptation by Flurin Darms), respectively. However, the Sutsilvan and Surmiran versions were still published in a joint edition. The first volume containing a joint edition of the Ladin and Sursilvan idioms was eventually published in 1980. This edition used Selina Chönz' Ladin version and Flurin Darms' Sursilvan version.

Lastly, although as already noted the books were not originally envisaged as a trilogy, there are several instances in the paratexts, especially the publisher's peritext, where the connection between the books is highlighted. There is no preface in either *Flurina* or *Der grosse Schnee*, which suggests that they are both a continuation of the preceding book. The title page of *Flurina* states that Flurina is Uorsin's sister: "Schellen-Urslis Schwester," which highlights this continuity. The front cover of *Der grosse Schnee* is also an illustration of Uorsin on skis carrying Flurina through a snowstorm. They both appear together at the center of the page, which is a contrast to the previous two books, where each character appeared on the front cover alone.

Repetition is also used to emphasize the chronological sequence between *Schellen-Ursli* and *Flurina und das Wildvöglein*. For example, the first three sketches in *Flurina und das Wildvöglein* mirror the first three illustrations in *Uorsin*. Some lines of *Flurina und das Wildvöglein*'s text also link the two books. Flurina is introduced with the same two lines that are used to introduce Uorsin in *Schellen-Ursli*: "Hoch in den Bergen, weit von hier, /da wohnt ein Mägdlein [or Bublein, in *Schellen-Ursli*'s case] so wie ihr" (Schultze-Kraft 163). In other words, Chönz repeats certain textual elements in *Flurina und das Wildvöglein* that have also appeared in *Schellen-Ursli*. According to Nikolajeva (197), children enjoy repetition, recognizability, and predictability since these features arouse the child's curiosity and therefore stimulate further reading. This is another form of textual transformation (Lefebvre) that is frequent in the creation of children's literature, and this adaptation is a way of prolonging the life—or afterlife (Benjamin)—of the work.

In addition, Carigiet maintains the style of the illustrations of *Uorsin* in *Flurina und das Wildvöglein* and *Der grosse Schnee* so they form a coherent whole. Except for the addition of small sketches alongside the main illustrations, the style remains the same, and several motifs are used throughout the trilogy. In *Uorsin*, the text and illustrations appear side by side: the text on the verso and the illustration on the recto, reflecting one another. Using Nikolajeva and Scott's typology for the wide diversity of word–image relationships in picturebooks, *Uorsin* and its two sequels fall into the symmetrical category. In other words, there is a mutually redundant nature of the interaction between the visual and verbal: "the words tell us exactly the same story as the one we can 'read' from the pictures" (Nikolajeva and Scott 14). Yet as the series progresses, there is an increasing use of illustration on the verso. In *Uorsin*, Carigiet's contribution remains confined on the recto, as does Chönz's on the verso. Yet in *Flurina und das Wildvöglein*, Carigiet's contribution spills over into the verso with small sketches, confined to the top of the page and depicting other short narratives of the text. In *Der grosse Schnee*, Carigiet's sketches on the verso take over the text and interact with the large illustrations on the recto (ten Doornkaat 68), thereby becoming the dominant feature of the book instead of acting as one among two equal elements of the iconotext. For example, in *Der grosse Schnee*, when Uorsin sets off on his skis to find Flurina, his line movement in the last sketch on the verso of him on skis continues straight down the fallen tree trunk depicted in the illustration on the recto. To the reader, it appears as if Uorsin is about to ski down the tree trunk, moving their gaze from the sketches on the verso to the illustration on the recto, where Uorsin finds a chain of lightbulbs to follow. The reader can follow this chain to find Flurina, as Uorsin does in the story. The chain in this illustration aligns with the location of the chain on the next double page spread, creating a continuity of movement over the page and over the double page spread to Flurina's location (Figs. 4 and 5).

The illustrations now dominate the text in the iconotext, with the textual element seemingly used as a verbal support. This shows a gradual shift from the text and illustrations having equal weight in the first book to the illustrations occupying more of the page, thus dominating the volume visually. The effect is to visibly uncrown or, in this instance, even marginalize the author from her own work. This is clear not only in the frequency and the placement of the illustrations but also in the paratexts. For example, on the title page of *Schellen-Ursli*, Chönz's name is listed

Fig. 4 Double page spread of Uorsin skiing down tree trunk towards the lights in *Der grosse Schnee*

Fig. 5 Uorsin following the chain of lights over the first double page spread onto the second to Flurina's location in *Der grosse Schnee*

before Carigiet's ("Erzählung: Selina Chönz Bilder: Alois Carigiet"). Yet in *Flurina und das Wildvöglein* and *Der grosse Schnee*, Carigiet's name precedes Chönz's and they no longer appear on the same line ("Bilder: Alois Carigiet / Erzählung: Selina Chönz"). On the one hand, this is due to the close interaction between images and text in picturebooks (O'Sullivan), but it could also be due to Carigiet's dominant position as the more well known of the two contributors in German-speaking Switzerland and is thus a way of drawing the consumer to buy the book. In comparison, the Romansh volumes of *Uorsin* and *Flurina* always place Chönz and Carigiet on the same line without defining their role. The Lia Rumantscha's insistence on maintaining both creators on the same line without defining their role is an attempt to give equal weight not only to both contributors but also to both idioms. It is only with the publishing of *La naivera* and *La cufla gronda* in 1964 that they privilege one contributor over the other. In the Putèr edition, Chönz is placed above Carigiet. This is because it will be sold in the region where she is from and is known. However, in the Sursilvan volume, this is reversed because Carigiet is from the Surselva and is more renowned there than Chönz. Of course, this is a marketing strategy, since locals from one region would have more of an affinity to one of the contributors than to the other and would therefore be more likely to buy the book if they see a name that is known to them.

The different contributors and their involvement in the production of the picturebooks raise questions regarding the singular ownership of the text (Nikolajeva and Scott 29). *Uorsin* is often attributed to Carigiet because, in comparison to the text, the images dominate the double page spread and because he was more famous than Chönz at the time of the book's first publication. Yet in the volume's paratext, the two are given equal weight. *Uorsin*'s cover, as we have seen, identifies both Chönz and Carigiet as authors, and their names are both depicted in equal size (not giving prominence to either). However, in all versions of the preface of both the German and Romansh editions (except for the 1971 Romansh preface), Pult mentions that it was Chönz who approached Carigiet with the story and not the other way around, as is often portrayed (Chönz, *Uorsin*, 1971, German edition). This shows that Pult is attempting to pinpoint the ownership of the text and the source of the initial inspiration. Yet it was neither the verbal nor the visual alone that resulted in *Uorsin*'s popularity: it was the iconotext as a whole, which is a product of all the contributors' involvement. However, since Carigiet was much more renowned at the time, Pult's persistence can be seen as an attempt to raise Chönz's subordinate status and bring the two contributors onto an equal footing—in other words, a dual ownership of equal weighting. Especially since Chönz's name is moved below Carigiet's on the title page in later editions of *Uorsin* and its sequels.

Conclusion

Chönz and Carigiet's work shows that translation can be an integral part of a picturebook produced in a minority language setting even before this is translated into another language. We have seen this, first, in the adaptation of Ladin into the other Romansh idioms and, second, in the intersemiotic translation of the text into illustrations and vice versa. The adaptations into the other Romansh idioms are eloquent textual representations of the multilayered nature of language and language politics surrounding minority language use, its promotion, and its relationship with the major language it is placed in relation with. In addition, the relationship between text and image brings questions surrounding authorship and originality to the fore. However, to give a full picture of this complex relationship between the verbal and visual, and of the process involved in translating a picturebook as a whole, a close textual analysis of the individual versions of *Uorsin* and its sequels is also needed to pinpoint the various ways in which the different contributors have had an influence in the development of the iconotext.

From the role reversal between illustrations and text to the adaptations into Romansh idioms of *Uorsin* and its sequels, we can see that neither translations nor languages are as binary as often suggested. The relationship between the so-called original and translation and between the target and source language is, in fact, much more fluid and complex. When translating a picturebook, not only may the illustrations and texts be translations of each other, but the unity of words and images is translated with the intent of producing (rewriting) a new iconotext (Oittinen, "On Translating" 110). If the process of translating picturebooks is taken in combination with the presence of a minority language, it can provide new insights into areas that are often overlooked when discussing the translation of picturebooks between major national languages—since questions regarding power relations, multilingualism, and authorship are more visible in such contexts. For this reason, more research needs to be done on the translation of children's literature in multilingual and minority language settings. Additionally, picturebooks are one of the key genres where text and image interact. Thus, research on picturebooks in minority languages can both broaden the definition of translation and provide answers to wider questions concerning translation, language, meaning, and the history of publishing practices. This line of investigation could also be enhanced through research into Xavier Koller's film adaptation of *Schellen-Ursli* (2015) and through further, in-depth analysis of what a shift from a Romansh-speaking Uorsin to a Swiss German-speaking Ursli means for all the issues and questions raised above.

Notes

1. Specifically for the German version.
2. Graubünden, also known as Grisons, is the largest and easternmost canton (or member state) of Switzerland.
3. Translated from the German *Idiom*.
4. The illustrations, Figs. 1–5 by Alois Carigiet are reproduced. *Coverillustration aus SCHELLEN-URSLI, Coverillustration aus FLURINA UND DAS WILDVÖGLEIN, und Coverillustration, Text und Innenillustrationen aus DER GROSSE SCHNEE von Alois Carigiet (Bild) und Selina Chönz (Text). Mit freundlicher Genehmigung des Orell Füssli Verlags ©1971 Orell Füssli Sicherheitsdruck AG, Zürich. Alle Rechte vorbehalten.*
5. I refer to these as joint editions and not parallel editions, since both idioms appear on the verso and the illustrations on the recto.
6. Emphasis in italics is my own.

7. Emphasis in italics is my own.
8. A situation where a standard variety is used in formal situations and a low variety is used in familiar and everyday situations.
9. The publisher of the German trilogy was the Schweizer Spiegel Verlag, which mainly published cultural history publications. The Lia Rumantscha, the umbrella association for Romansh language and culture, published the Romansh editions.

Works Cited

Children's Books

Chönz, Selina. *Schellen-Ursli*. Zurich: Schweizer Spiegel Verlag, 1945.

———. *Uorsin*. Translated by Curò Mani (Sutsilvan) and Alexander Lozza (Surmiran). Chur: Lia Rumantscha, 1945. (Surmiran/Sutsilvan)

———. *Uorsin*. Translated by Gion Cadieli (Sursilvan). Chur: Lia Rumantscha, 1945. (Ladin/Sursilvan)

———. *Flurina und das Wildvöglein*. Zurich: Schweizer Spiegel Verlag, 1952.

———. *Florina and the Wild Bird*. Translated by Anne and Ian Serraillier. London: Oxford University Press, 1953. (English)

———. *Flurina e gl'utschelet salvadi/Flurina a gl'utschiet salvadi*. Translated by Anna Capadrutt (Sutsilvan) and Gion Peder Thöny (Surmiran). Chur: Lia Rumantscha, 1953. (Surmiran/Sutsilvan)

———. *Flurina e l'utscheïn sulvedi/Flurina e gl'utschiet selvadi*. Translated by Decurtins, Alex (Sursilvan). Chur: Lia Rumantscha, 1953. (Ladin/Sursilvan)

———. *Arupusu no Kyodai*. Translated by Unknown. Tokyo: Iwanami-shoten, 1954. (Japanses dual edition containing *Uorsin* and *Flurina*)

———. *Catherine et l'Oiseau sauvage*. Translated by Maurice Zermatten. Bruges: Desclée de Brouwer, 1955. (French).

———. *Der grosse Schnee*. Zurich: Schweizer Spiegel Verlag, 1955.

———. *La grande neige*. Translated by Blaise Briod. Bruges: Desclée de Brouwer, 1956. (French)

———. *The Snowstorm*. Translated by Anne and Ian Serraillier. London: Oxford University Press, 1961. (English)

———. *La cuffla gronda*. Translated by Flurin Darms. Chur: Lia Rumantscha, 1964. (Sursilvan)

———. *La naivera*. Chur: Lia Rumantscha, 1964. (Ladin)

———. *Schellen-Ursli*. Zurich: Orell Füssli Verlag, 1971.

———. *Uorsin*. Translated by Gion Cadieli. Chur: Lia Rumantscha, 1971. (Sursilvan)

———. *La naivera/La cuffla gronda*. Translated by Flurin Darms (Sursilvan). Chur: Lia Rumantscha, 1980. (Ladin/Surslivan)

Secondary Sources

Bakhtin, Mikhail [Bahtin, Mihail]. *Problems of Dostoyevsky's Poetics* (orig. *Problemy poetiki Dostoevskogo* 1963). Translated and edited by Caryl Emerson. University of Minnesota, 1987.

Benjamin, Walter. "The Translator's Task." Translated by Steven Rendall. *TTR: Traduction, Terminologie, Rédaction*, vol. 10, no. 2, 1997, pp. 151–165.

Berg-Ehlers, Luise. *Berühmte Kinderbuchautorinnen und ihre Heldinnen und Helden*. Munich: Elisabeth Sandmann Verlag Gmbh, 2017, pp. 42–45.

Brodzki, Bella. *Can These Bones Live? Translation, Survival, and Cultural Memory*. Stanford University Press, 2007.

Carigiet, Alois. "Auf der Suche nach dem verlorenen Paradies der Kindheit." *Die Weltwoche* dated 9 December 1966.

Cathomas, Bernhard. "Rhaeto-Romansh in Switzerland up to 1940." *Ethnic Groups and Language Rights*, edited by Sergiji Vilfan. Aldershot: Dartmouth Publishing Company Ltd., 1993.

Deplazes, Gion. "Litteratura Rumantscha." *Lexikon Istoric Retic*. Undated.

Genette, Gérard. *Paratexts: Thresholds of Interpretation*. Translated by Jane E. Lewin. Cambridge University Press, 1997.

Hofmann Fadrina. *Was Jon Pult zu sagen hatte, ist noch heute relevant*. Südostschweiz, 2011.

Hutcheon, Linda. *A Theory of Adaptation*. Routledge, 2006.

Jakobson, Roman. "On Linguistic Aspects of Translation." *The Translation Studies Reader*, edited by Lawrence Venuti. Routledge, 1959, pp. 232–239.

Lathey, Gillian. *Translating Children's Literature*. Routledge, 2016.

Lefevere, André. *Translation, Rewriting, and the Manipulation of Literary Fame*. Routledge, 1992.

Lefebvre, Benjamin. "Introduction." *Textual Transformations in Children's Literature: Adaptations, Translation, Reconsiderations*, edited by Benjamin Lefebvre. Routledge, 2013, pp. 1–7.

Kaindl, Klaus. "Multimodality and Translation." *The Routledge Handbook of Translation Studies*, edited by Carmen Millán and Francesca Bartrina. Routledge, 2013, pp. 257–270.

Krättli, Esther. "Curò Mani." *Theaterlexikon der Schweiz*, vol. 2, edited by Andreas Kotte. Zurich: Chronos Verlag, 2005, pp. 1167–1168.

Meylaerts, Reine. "Heterolingualism in/and Translation: How legitimate is the Other and his/her Language? An introduction." *Target*, vol. 18, no. 1, 2006, pp. 1–15.

Müller, Rafael. *Pader Alexander Lozza—il pader stravagant*. RTR, 2015.

Neue Zürcher Zeitung. "50 Jahre «Schellenursli»—kühne Graphik und Ursprungsmythos," no. 56, 1995.

Nikolajeva, Maria. "Beyond Happily Ever After: The Aesthetic Dilemma of Multivolume Fiction for Children." *Textual Transformations in Children's*

Literature: Adaptations, Translation, Reconsiderations, edited by Benjamin Lefebvre. Routledge, 2013, pp. 197–213.

Nikolajeva, Maria and Carole Scott. *How Picturebooks Work*. Routledge, 2001.

Oittinen, Riitta. *Translating for Children*. Garland Publishing, Inc., 2000.

———. "On Translating Picture Books." *Perspectives*, vol. 9, no. 2, 2001, pp. 109–125, https://doi.org/10.1080/0907676x.2001.9961410.

———. "Where the Wild Things Are: Translating Picture Books." *Meta*, vol. 48, no. 1–2, 2003, pp. 128–141, https://doi.org/10.7202/006962ar.

———. "The Verbal and the Visual: On the Carnivalism and Dialogics of Translating for Children." *The Translation of Children's Literature: A reader*, edited by Gillian Lathey. Clevedon: Multilingual Matters Ltd, 2006, pp. 84–97.

O'Sullivan, Emer. "Translating Pictures: The Interaction of Pictures and Words in the Translation of Picture Books." *European Children's Literature II*, edited by P. Cotton. Kingston University, 1998, pp. 109–120.

Pult, Chasper. "Ein Engadiner Kuppler als Geburtshelfer des Uorsin: Die Rolle von Jon Pult bei der Entsteheng des Schellen-Ursli." *Alois Carigiet—Kunst, Grafik, Schellen-Ursli*, edited by Hans ten Doornkaat. Zurich: Orell Füssli Verlag, 2015, pp. 40–43.

———. "Kastanien mit Schlagrahm oder Milchreis? Der eigenwillige Umgang mit dem Original in den Übersetzungen des Uorsin." *Alois Carigiet Kunst, Grafik, Schellen-Ursli*, edited by Hans ten Doornkaat. Zurich: Orell Füssli Verlag, 2015, pp. 50–55.

ten Doornkaat, Hans. "Einzelbilder und doch Bildfolgen: Carigiets Umgang mit dem Medium Bilderbuch." *Alois Carigiet—Kunst, Grafik, Schellen-Ursli*, edited by Hans ten Doornkaat. Zurich: Orell Füssli Verlag, 2015, pp. 66–69.

Trullman, David. *Uorsin Superstar*. RTR, 2007.

Schultze-Kraft, Ofelia. "Flurina und Schellen-Ursli. Die «Engadiner-Trilogie» von Selina Chönz und Alois Carigiet." *Dichterische Freiheit und pädagogische Utopie: Studien zur schweizerischen Jugendliteratur*, edited by Heidy Margrit Müller. Bern: Lang, 1998, pp. 158–186.

Studer, Thomas. "Gusti Güggel Alias Franz von Hahn: Zur Sprache von schweizer Kinderbüchern vor dem Hintergrund der deutschschweizer Sprachsituation." *Nebenan: Der Anteil der Schweiz an der deutschsprachigen Kinder- und Jugendliteratur*, edited by Verena Rutschmann. Zurich: Chronos Verlag, 1999, pp. 79–104.

Valär, Rico. "Die Inszenierung als uralte, wurzelstarke, urschweizerische Bergsprache« : Alois Carigiet und die rätoromanische Heimatbewegung." *Alois Carigiet—Kunst, Grafik, Schellen-Ursli*, edited by Hans ten Doornkaat. Zurich: Orell Füssli Verlag, 2015, pp. 36–39.

Mixing Moralizing with Enfreakment: Polish-Language Rewritings of Heinrich Hoffmann's Classic *Struwwelpeter* (1845)

Joanna Dybiec-Gajer

From Invisibility to Celebrity—An Overview of Polish-Language Editions

The history of Polish-language *Struwwelpeter* began in St. Petersburg—where in 1858 the Warsaw-born publisher and editor Bolesław Maurycy Wolff, later referred to as the tsar of books, published *Złota różdżka* [A Golden Wand], a rewriting and cultural adaptation of an earlier Russian edition (Fig. 1).[1] As elsewhere in Europe, the book proved to be a success, and new editions, as well as "struwwelpetriades" or spin-offs, followed. However, its popularity was not as widespread as, for example, in the Anglophone world, since Polish literature, for children and adults alike, remained wary of nonsense, black humor, and the macabre.

The publication and reception history of *Złota różdżka* is characterized by a double invisibility. The book appeared anonymously: Neither the author's nor the translator's name was specified. A publishing practice of failing to give credit to the translator is not surprising in the context of early modern literature for children. In all *Struwwelpeter* translations in England (1848–1910), the translators were unnamed (O'Sullivan,

J. Dybiec-Gajer (✉)
Pedagogical University of Krakow, Krakow, Poland

A. Kérchy and B. Sundmark (eds.), *Translating and Transmediating Children's Literature*, Critical Approaches to Children's Literature,
https://doi.org/10.1007/978-3-030-52527-9_4

Fig. 1 Struwwelpeter as Stepka Rastrepka (Shock-headed Steve) from the Russian edition (1849). The images from this edition were re-printed with slight modifications by first Polish-language publications of Hoffmann's classic

"Anything to me is sweeter" 60), and they have remained unidentified despite academic scrutiny and effort.[2] Such anonymity was also not exceptional in other settings of literary translation, which is well documented in the subject literature, providing ample evidence for Venuti's now classic concept of "the translator's invisibility" (1995). As for the author's identity, writing for children was not considered socially prestigious or ennobling, which is reflected in Heinrich Hoffmann's decision to hide behind a penname of Reimerich Kinderlieb; it was only in 1847 that his real name was disclosed for the first time in a *Struwwelpeter* edition. Moreover, as part of the marketing strategy of the first Polish publisher, *Złota różdżka*, like its Russian predecessor, swerved in its outward appearances from the German text published as *Lustige Geschichten und drollige Bilder* [Funny Stories and Humorous Pictures]. Both textual and illustrative domestication or, to use an anachronistic concept, glocalization contributed to its success. The tradition of domestication, a continued peripheral status of writing for children, and a disregard for German children's literature after World War II had their consequences in that Heinrich Hoffmann as the author was first featured on the book cover as late as 2003. It was in the same publication, a facsimile of a Wolff edition prepared by Junusz Dunin (2003), that a tentative attribution of the translation to Wacław Szymanowski was discussed in some detail.

Further, another characteristic and defining feature of the Polish *Struwwelpeter* is the long-lasting textual stability of the first target text. Despite a number of editions documenting the continued popularity of the book until World War II, the text version of 1858 remained unchanged in circulation. Even the re-illustration prepared by Bogdan Szymanowski in the interwar period had the very same text at its core (1922, 1933). Interestingly, this was not taken as an opportunity to redress some incongruities between the text and the image. Hoffmann's soup-defying Kaspar dies on the fifth day, likewise, his Russian counterpart Fryc. However, for the sake of rhyme, the Polish Michaś has his life extended by one day although this stands in opposition to the illustrations of Wolff's publications, where the boy's grave is clearly marked by a large "5." The same setup is repeated in the re-illustrated edition (Fig. 2).

The domination of the first rewriting along with its textual stability were challenged almost 160 years after the first market appearance of *Złota różdżka* and then almost concurrently by a host of three publications, two of which have a direct bearing on this paper as children's books and will be discussed here.[3] In 2015, a rendition by Lech Konopiński appeared in Edition Tintenfaß (Germany) to be the first Polish-language edition to feature Hoffmann's original illustrations and to include all ten stories of his *Struwwelpeter* in its canonical form (as the previous publications did not include "Vorspruch" and "Die Geschichte vom wilden Jäger"). Praised in the publisher's foreword as providing a translation equal in its qualities to the merits of the original, Lech Konopiński, who can be legitimately considered as a translator and literary figure of some standing, provides a translation that is one-man's work. A different

Fig. 2 Struwwelpeter as Staś Straszydło (Frightening Stan) from the Polish re-illustrated edition (1922). Cover detail (ill. Bogdan Nowakowski)

concept follows in 2017 by Egmont, a media conglomerate, which in its "Art" series brought out a both textually and graphically modernized and reconceptualized *Złota różdżka*. Its publication is a collaborative effort of six rather well-known literary translators who have gained reputation for their work and established themselves in the field. Having Michał Rusinek in the project—a secretary of the Nobel Prize Winner for Literature, Wisława Szymborska, and a writer and translator of children's books himself—adds a touch of literary celebrity. Since the book is designed as an adaptation, bringing together a team of experienced translators and letting them free their imagination, it offers a creative, novel, and promising approach to the source text. This is also used as a marketing strategy, as the names of literary celebrities or celebrities-to-be are featured in various paratexts and promotional materials as "brands" (cf. Szwebs).

The First Translation—Humor, Horror, and Moralization

The first translation, with its distinct narratorial voice, is of interest here as a starting point and background for understanding and interpreting subsequent rewritings. Its long presence in the Polish culture has influenced *Struwwelpeter*'s reception and standing in the literary polysystem[4]; it is also evidence of the lasting quality and appeal of *Złota różdżka*. The translator succeeded in rendering the dynamism, vitality, and vibrancy as well as the humor characteristic of both Hoffmann's text and its Russian retelling. It is rhythmic and well-suited for out-loud reading. What may come as a surprise to *Struwwelpeter* readers is that *Złota różdżka* is remembered and present in the subject literature and popular lore mostly by means of its catchy moralizing punch lines. But where are these in Hoffmann's text or in its Russian rewriting?

In *Struwwelpeter*, the narrator is decidedly and consistently withdrawn, refraining from making comments, judgments, or expressing emotions. In her study of British translations, Emer O'Sullivan notes, "One of the most striking elements of Heinrich Hoffmann's (original) *Struwwelpeter*, and one that explains the different critical responses to the book, is the attitude of the narrator toward his young protagonists" ("Anything" 67). The narrator's stance seems ambivalent or even controversial since he "fails to condemn his young protagonists outright for their behavior" (68). Likewise, the construction of the narrator's voice is such so as not to direct the reader's sympathies by evaluating children's behavior or by giving insights into the feelings of the punished protagonist. The exceptions are scarce—for example, when Paulinchen is called "das arme Kind" ("poor child") or when thumbless Konrad looks sad ("sieht der Konrad traurig aus").

The narrator of the first Polish translation, unlike that of the German and Russian source texts, is quick to insert moralizing comments or punch lines: for example, "Oj, od złości strzeżcie się, /Bo złym zawsze bywa źle" ("Oh, beware of anger, because the angry and the evil fare bad") in "Zły Józio," a retelling of "Die Geschichte vom bösen Friederich." In Polish, the adjective *zły* means both "angry" and "evil," the same as the German *böse*. More than half of the translated tales have moralizing conclusions added. Thus, the ambivalent stance of the original narrator is placed within the frames of the more conventional, or clichéd, cautionary tale with explicit rhyming morals. Interestingly, not only is the narrator of the Polish text a moralist, but he also suffuses the tales with an intriguing mixture of humor and horror.

Some of the punishment that befalls Hoffmann's protagonists is hyperbolized, for example, Little Suck-a-Thumb, named Juleczek, not only has his thumbs cut off as in the original but is also reprimanded by the tailor, who is silent in the German version, and on top of that, refused cakes by his mother. The most radical shift, at odds with Hoffmann's program of parents and caretakers consistently refraining from any punishment, is the severe beating meted out to Pawełek (Fidgety Philip). The heightened horror is in turn subverted through humor achieved by quick-paced rhymes and creative additions. For instance, Józio (Frederick) is also presented as a destroyer of books: "darł książeczki/W kawałeczki," although this amplification is not dictated by formal constraints of rhyme or prosody. Further, in the Polish rendition of the opening poem, "Struwwelpeter," the protagonist is named by making a direct reference not to his unkempt appearance (as in the German original, Russian Stepka Rastrepka or Twain's Slovenly Peter) but to his fear evoking capacity: Staś Straszydło (lit. Frightening Stan). Yet when compared with the illustration depicting a mildly ashamed, downward-looking boy, the eponymous Staś loses his fear evoking quality.

The first translation offers a creative rewriting, enhancing *Struwwelpeter*'s original qualities. Penned in the mid-nineteenth century, it grows out of a translation and publication culture that was less concerned with faithfulness or copyrights[5] than the contemporary times. The subsequent retelling, almost 160 years later, is a product of different times and speaks with a different narratorial voice.

The Second Translation—Fidelity and Diminutives

An analysis of translations' paratextual framework can provide interesting insights into production, distribution, and interpretation of translated texts (cf. Tahir-Gürçağlar). The title of the second Polish translation (in 2015) is an important paratext as it serves as a threshold for reading and interpreting. In different terminology, the title is a "critical point" (cf. Dybiec-Gajer, "The Challenge of Simplicity"): The way in which it is transferred into a new linguistic and cultural context has significant implications for the functioning of the entire target text. A close reading of the title—*Piotruś Rozczochraniec. Wesołe historyjki i zabawne rysunki*—clearly reveals the translator's programmatic strategy, based (in my opinion) on the tenets of fidelity toward the source text and child-directed speech. It is the only Polish *Struwwelpeter* edition that does not change or abridge the title and that abandons "the golden wand" tradition, following verbatim the German wording. In the first translation, Hoffmann's "humorous stories and funny pictures" were replaced by a rhymed didactic instruction "czytajcie dzieci, uczcie się, jak to niegrzecznym bywa źle," literally "children, read and learn how badly the naughty fare." In Konopiński's rendition, *Struwwelpeter* faithfully becomes *Piotruś Rozczochraniec* (Shock-headed Peter). Again, this shows, on the one hand, an orientation toward the source text and, on the other—by giving the protagonist name in a diminutive form (Piotruś) rather than in a full and standard one (Piotr)—an orientation toward the usage of language imitating to some extent child-directed speech (CDS).

In brief, the programmatic source-text orientation of this publication in both word and image can be summarized in the following points:

- completeness (All ten tales and the introduction of the original are included).
- usage of Hoffmann's original illustrations.
- preservation of standard proper names in their original form, sometimes with small phonetic adjustments (With a few exceptions such as the eponymous Piotruś for Peter or the replacement of Hans with its Polish equivalent, Janek).
- careful use of amplification (Unlike the first translator, who freely added the narrator's comments and extended the length of the translation, Konopiński sparsely uses this technique).
- formal correspondence (The majority of the translated poems have the same number of verses).
- accurateness in transferring information content, with its extent varying from poem to poem.

To illustrate how fidelity is at play with reference to the transfer of informational load, let us analyze the final scene of "Die Geschichte vom bösen Friederich." In the history of *Struwwelpeter*'s translations, it has been frequently subject to cultural adaptation. The original ends with the narrator specifying the dog's menu: cake, leberwurst, and wine—all of which can be seen in the picture accompanying this topsy-turvy setting. In Victorian England wine was believed to be an unsuitable item for a children's book so in early British translations the dog enjoys only "soup, pies and puddings" (1848) or simply "beef" (1900) (O'Sullivan, "Anything to me is sweeter" 63). By contrast, Mark Twain, following the exaggeration of the tall tale tradition, considerably extends Hoffmann's brief description. He presents the dog as a food connoisseur, who "[s]ings praises soft and sweet and low" over the boy's dinner, and "sips the wine, so rich and red, /And feels it swimming in his head". Wine is missing in the Russian *Stepka*, so it comes as no surprise that the dog just eats "obiad" ("dinner") in the first Polish translation. In Konopiński's matter-of-fact rendition of this scene, all the items of the original menu are faithfully preserved, with the exception of "schab" ("pork"), which is added for the sake of rhyme.

Przy stoliczku Fryca	
słodką babkę będzie jeść;	[sweet babka cake]
wątrobiankę zje i **schab**,	[liver sausage, pork]
nie wypuści **winka** z łap.	[wine]

Interestingly, the source text's generic cake ("Kuchen") in the Polish translation becomes a specific type of a cake, a traditional festive "babka," baked for Easter holidays. In this way, the translator's explicitation brings his translation much closer to Hoffmann's drawing than the author's own verse, because the image shows a characteristic high and round "babka" or "Bundt cake."

As for CDS (child-directed speech, a term used in child development studies, or colloquially "baby talk"), it is defined as a type of speech that adults tend to use in addressing infants and children that is different from inter-adult communication. Transposed to the field of literature, this generally means a simplification of language to make it more accessible to children. One of CDS's salient features involves the frequent use of diminutives, which in inflectional languages such as Polish is especially productive. The rendition of "Die Geschichte vom bösen Friederich" shows pronouncedly the implications of CDS usage for the source text-oriented strategy of the publication. It is in this poem, exceptionally, that Hoffmann's distanced narrator uses evaluative language, calling the protagonist "bitterbös" ("wicked") and "arger Wütterich" ("terrible brute"). In the translation, the child's wickedness is severely undermined by a repeated and unmotivated use of diminutives. Choosing the diminutive form of the name for the title of the poem ("Dzieje Fryderyczka złośniczka") is not conditioned by linguistic constraints, as the basic form would also lead to a rhyme (Dzieje Fryderyka złośnika). Moreover, the opening of the poem reinforces the effect of belittling the boy's bad behavior by calling him Fryderyczek, Frycek, Fryc, which are diminutivized forms of the protagonist's name.

The use of CDS also undermines the original message in the rendition of the eponymous "Struwwelpeter." Let us begin with a short analysis of the narrative voice of the original:

> The unkempt, pedestaled exhibitionist Struwwelpeter is described by Hoffmann as a boy who will not let his nails be cut or his hair combed. The first negative expletive in the verse, "Pfui" ("yuck"), is expressed by the narrator himself. Its repetition, together with the adjective "garstig" ("horrid")—"Pfui, ruft da ein jeder: /Garstger Struwwelpeter!"—is *society's* comment and is merely related by the narrator. (O'Sullivan, "Anything" 68)

The Polish narrator does not express any negative emotions such as disgust or condemnation. He uses CDS in an interactionist dimension, addressing the child listener directly as "kochanie" ("darling"). Moreover, he uses diminutives not only for the protagonist's name but also for his body parts (instead of "na rękach" he prefers "na rączkach," meaning "on little hands"). Such a protagonist is no longer a repellent sloven or a frightening scarecrow but an amiable little boy with unkempt hair and long nails. Further, such a description is also at odds with the picture, featuring a boy figure far too old to have "little hands." The translation also fails to provide the reason for the boy's appearance as the reader does not learn about Struwwelpeter/Piotruś's stubborn refusal to have his hair cut and nails trimmed. Therefore, the didactic potential of the original text is lost.

The Adaptation—Fragmentation and Enfreakment

The latest publication of *Złota różdżka* is different from the previous one in that it is designed not as a translation but as a deliberate adaptation and re-illustration, meant to breathe a new life into a somewhat forgotten classic. Given the fact that part of the design was to distribute texts among a number of translators, the book by the nature of its creation makes a fragmented picture. There seem to be at least as many narrative voices as there are translators participating in the adaptation. Approaching the individual poems on their own and as non-translations from a purely formalistic side, any fair critic could not fail to notice a certain excellence of linguistic expression and creativity. Approaching the poems as related to a source text and taking as a criterion the orientation toward the original narrator's withdrawn stance and lack of verbalized moralizing, the tales can be grouped into three categories:

1. Modernizing adaptations: Zuzanna Naczyńska's "Opowieść o małych czarnych chłopcach" [The tale of little black boys] and "Opowieść o Jacku Wniebogapku" [The tale of Jacek stare-in-the-air]
2. Adaptations of *Złota różdżka*: Karolina Iwaszkiewicz's "O złym Szymku" [The angry-evil Szymek], "O Filipie, co się bujał" [Fidgety Filip] and Marcin Wróbel's "O Konradzie, co obgryzał paznokcie" [Konrad who would bite his nails]
3. Re-translations: The remaining poems

The modernizing adaptations go beyond a mere linguistic update of the vocabulary and transfer the tales to the contemporary settings with their new problems and concerns. In "The tale of little black boys," condemning the children's mockery of a different looking boy, the young protagonists—who have received popular names of Janek, Franek, and Kuba—are equipped with contemporary attributes, such as a scooter, roller skates, and a mobile, which supersede Hoffmann's "Brezel" ("brezel"), "reif" ("circle"), and "Fänchen" ("little flag"). Their form of mocking is also brought up to date, as Kuba films the scene with his mobile phone. Moreover, unlike the protagonists of the original or the previous Polish translations, who only insulted the black boy verbally, here "the new kids from the block" resort to physical violence, as one of them throws a stone. The punishing agent—in the original, Nikolas, and in the first Polish translation, an old magician, here, a janitor ("dozorca")—puts on magical clothes and uses a wand to make the boys jump into a puddle. In this way, a link is made both to the contemporary meaning of the eponymous "różdżka" ("wand") as a sorcerer's attribute and to the first Polish translation. The final word is given to the black boy, who explains that he is not really a stranger as he was born in Poland. A large Polish flag reinforces this specifically national context. Thus, the more universal condemnation of mocking blackness as otherness that is present in Hoffmann's original text or previous Polish translations is here more pronouncedly addressed at Polish readers. In the second adaption, "The tale of Jacek stare-in-the-air") fewer changes have been introduced to the original concept. The protagonist, like Hoffmann's Hans, stares in the air rather than in the mobile as in some contemporary German rewritings and pastiches, where Hans Guck-in-die-Luft becomes Hans Guck-in-die App. The most important change is that the river, into which the boy falls, is heavily polluted because it contains the city's sewage. As a result, Jacek not only gets wet and laughed out by fish as in the original but also gets a rash. Such increasing of the punishment is in line with the tradition of the first translation.

While the modernizing adaptations seem to follow the tenets of littérature engagée, attempting to highlight the problems of racial discrimination and environmental pollution, the motivation and the overall goal of the adaptations, which are clearly based not on the original but on the first Polish retelling, remain opaque. "The tale of the Fidgety Filip" ("O Filipie, co się bujał"), follows in the footsteps of the first translator, who introduced corporal punishment, and depicts a family scene at the dinner table in which a father insults a fiddling boy in an attempt to make him eat his dinner. The scene ends in a little family disaster, with dinner on the floor and the father severely beating his son. The adaptation transforms the first Polish translation into a portrayal of the near pathological and sadistic child–parent relations, yet the purpose of this literary indulgence in violence does not become apparent. In contrast, Mark Twain's free translation highlights different qualities of the protagonist's play, and the father, as in the original, refrains from any punishment.

The last category of re-translations is the most heterogeneous with reference to the narrator's voice. For the sake of the clarity of the presentation, it can be generalized that the narrator, especially in comparison with the first translation, is considerably withdrawn. An analysis of the opening "Struwwelpeter" brings us to a second important characteristic feature of the discussed edition—enfreakment.

In the tale's verbal layer, Piotruś Czupiradło—literally, "Shock-headed Piotruś"—is depicted as a "creature" ("stwór") that is wary of a comb and scissors. Consequently, "his nails, black as sooth, are more than a meter long" ("Paznokcie czarne jak sadza mają już metr—nie przesadzam"). Yet in the illustrative layer, it is not the unkempt qualities that are characteristic of the protagonist but the fact that it is not a human creature but a curiosity-evoking monster (see Fig. 3) (cf. Dybiec-Gajer 2020).

Fig. 3 Struwwelpeter as Piotruś Czupiradło (Shock-headed Peter) (2017) (ill. Justyna Sokołowska). One of three versions of the protagonist. Image courtesy of the artist

It is presented in three different basic versions: one-eyed, devil-like and three-eyed with a vampire's teeth and octopus-like legs. Cut in half, the monster is printed horizontally on as many as six double spreads, allowing for different configurations of the images. They correspond with the poem which is constructed as a puzzle addressed at the reader whose aim is to guess the identity of the protagonist. It is telling that the monster's appearance is to some extent reminiscent of a figure chosen for the cover of a monograph dedicated to the cultural history of freak shows (Kerchy, Zitllau 2012). It is not a coinsidence since the monster can be aptly described as a fragmented and freakish construction of contemporary popular culture.

Given the variety of voices in the discussed edition of *Złota różdżka*, it is difficult to generalize or draw conclusions. However, if we focus on the qualities of the narrator, it will become apparent that he is not as withdrawn as in the original and that some verbally expressed didacticism seems to be present also in the latest edition. Thus, Szymek (cruel Frederick) will not hit anyone because the dog is taking care of his whip, and Konrad, in a macabre rewriting in which he has his arms cut off, promises to leave in peace his toenails. Further, reflections of the didactic, moral pointing finger can be found in the zeal of modernizing adaptations reminiscent of the engaged literature.

Conclusion

The ambivalence of the source-text narrator, who does not take a stance on the actions of the protagonists or give any explicit prospect of the protagonists' improvement, seems psychologically and pedagogically difficult to accept across different cultures and times. Analyzing the Polish rewritings of the classic, one can discern various translation strategies and techniques that can be interpreted as compensating for or filling in the void of the German original. The first rewriting (in 1858) is a testimony to the nineteenth-century translation culture, which allowed mediators larger liberties in their treatment of source texts; it creates an opinionated, convivial, and quick-witted narrator who provides didactic and moralizing sayings and punch lines with ease and gusto, most of which are not motivated by the mediation of a third language, Russian. The second rewriting (in 2015), designed as translation, can be placed in the paradigm of equivalence. On the one hand, it manages to convey numerous aspects omitted or changed in the previous Polish editions. On the other hand, its frequent usage of diminutives, not always motivated, sometimes undermines the message of the tales. The narrator, generally withdrawn, occasionally slips in a comment or two, bringing the texts closer to horizons of conventional expectations about children's literature in the Polish polysystem. The third rewriting (in 2017) opens an entirely new chapter in *Struwwelpeter*'s Polish journey since it is not only a deliberate textual adaptation but also a re-illustration. At first sight, it is the illustrations that carry the innovative load of the edition. Playful, colorful, large-scale pictures tending to contemporary tastes include postmodern play with macabre and death, while the main protagonist Struwwelpeter is transformed from a long-haired and long-nailed boy into a fragmented pink freak of many guises. It is interesting that in such a modern and vibrant rewriting, the echoes of didacticism and moralizing can still be heard—either in the numerous voices that seem to correspond to the multiple narrators or in the stance of littérature engagée. Readers, parents, and critics may love or hate the new adaptation, but it is beyond doubt that *Struwwelpeter* and *Złota różdżka* live on and continue to generate new meanings and controversies in times of apparently no limits and taboos.

Notes

1. The source text for the first Polish rewriting of *Struwwelpeter* was *Стёпка-Растрёпка* (1849). The Russian title has been transcribed in a number of ways in English and other Latin-based alphabets (e.g. as Styopka Rastryopka), here the transcription Stepka Rastrepka is followed. The title of the Polish rewriting, *Złota różdżka*, can be translated as a "golden wand" if the predominant contemporary understanding of the word "różdżka" is used or as a "golden rod" or "birch," that is an instrument of punishment, if the historical approach is taken.
2. The English translator has been identified, yet inconclusively, to be Alexander Platt. See Brown and Jones, "*The English Struwwelpeter* and the Birth of International Copyright."
3. The third publication (Dybiec-Gajer, *Złota różdżka*) is not a children's book but a critical edition from the perspective of translation studies research.
4. For a number of reasons *Złota różdżka* has not attracted much serious academic attention in Polish research. It has frequently been subject to simplifying psychoanalytic analysis, in which it has been taken to be identical with the original (e.g., in Slany).
5. At that time, the Russian empire was not part of copyrights agreements. For more on copyrights and *Struwwelpeter's* history in the English speaking world, see Brown and Jones, "*The English Struwwelpeter.*"

Works Cited

Source Texts

Hoffmann, Heinrich. *Stepka Rastrepka. Rasskazy dlya detei*. St. Petersburg: Ginze [Hintze], 1849.

———. *Złota różdżka—Staś straszydło. Czytajcie dzieci, uczcie się, jak to niegrzecznym bywa źle*. St. Petersburg: B. M. Wolff, 1892.

———. *Złota różdżka—Staś straszydło. Czytajcie dzieci, uczcie się, jak to niegrzecznym bywa źle*. Illustrated by Bogdan Nowakowski. Warszawa, 1922.

———. *Der Struwwelpeter oder lustige Geschichten und drollige Bilder für Kinder von 3–6 Jahren von Dr. Heinrich Hoffmann. 585. Auflage*. Frankfurt am Main: Rütten und Loening, 1929.

———. *Slovenly Peter [Der Struwwelpeter]. Translated into English jingles from the original German of Dr. Heinrich Hoffmann by Mark Twain with Dr. Hoffmann's illustrations adapted from the rare first edition by Fritz Kredel*. New York: Limited Editions Club, 1935.

———. *Piotruś Rozczochraniec. Wesołe historyjki i zabawne rysunki*. Translated by Lech Konopiński, Neckarsteinach: Edition Tintenfass, 2015.

———. *Złota różdżka, czyli bajki dla niegrzecznych dzieci. Translated by Anna Bańkowska, Karolina Iwaszkiewicz, Zuzanna Naczyńska, Adam Pluszka, Michał Rusinek and Marcin Wróbel*. Illustrated by Justyna Sokołowska. Warszawa: Egmont Art, 2017.

Secondary Texts

Brown, Jane and Gregory Jones. "*The English Struwwelpeter* and the Birth of International Copyright." *Library*, vol. 14, no. 4, 2013, pp. 383–427.

———. "Who translated *The English Struwwelpeter*? The Self-Effacing Alexander Platt." *Struwwelpost*, vol. 21, 2015, pp. 20–24.

Dunin, Janusz. Złota Rószczka. Reedycja petersburskiego wydania z roku 1883/Heinrich Hoffmann. Struwwelpeter, Stepka-Rastrepka czyli Złota różdżka. Z dziejów kariery jednej książki/Janusz Dunin. Łódź: Verso, 2003.

Dybiec-Gajer, Joanna. "The Challenge of Simplicity: *Le Petit Prince* in Polish and English Translation from the Perspective of Critical Point Analysis." *Le Petit Prince et les amis au pays des traductions. Études dédiées à Urszula Dąmbska-Prokop*, edited by J. Górnikiewicz, I. Piechnik, and M. Świątkowska. Kraków: Księgarnia Akademicka, 2012, pp. 20–27.

———. "Von der Popularität in die Vergessenheit. Rätselhafte Wege des polnischen Struwwelpeter—zum 170. Jubiläum der Erstausgabe." *Struwwelpost*, vol. 22, 2016, pp. 4–13.

———. *Złota różdżka. Od książki dla dzieci po dreszczowiec raczej dla dorosłych*. Kraków: Tertium, 2017.

———. "Postanthropocentric Transformations in Children's Literature: Transcreating *Struwwelpeter*." *Negotiating Translation and Transcreation of Children's Literature*, edited by J. Dybiec-Gajer, R. Oittinen, and M. Kodura, Springer, 2020, pp. 39–55.

Ewers, Hans-Heino. "Die Schlingel hat die Welt erobert, ganz friedlich, ohne Blutvergießen. Warum der Struwwelpeter bis heute ein Bestseller ist." *Forschung Frankfurt*, vol. 1, 2009.

Kérchy, Anna and Andrea Zittlau, editors. *Exploring the Cultural History of Continental European Freak Shows and 'Enfreakment'*. Cambridge Scholars Publishing, 2012.

O'Sullivan, Emer. "Anything to Me is Sweeter… British Translations of Heinrich Hoffmann's *Struwwelpeter.*" *Princeton University Library Chronicle*, vol. 62, no. 1, 2000, pp. 59–71.

———. *Kinderliterarische Komparatistik.* Heidelberg: Universitätsverlag C. Winter, 2000.

Rühle, Reiner. "*Böse Kinder*": *Kommentierte Bibliographie von Struwwelpetriaden und Max-und-Moritziaden mit biographischen Daten zu Verfassern und Illustration.* Osnabrück: Wenner, 1999.

Sauer, Walter. *Der Struwwelpeter und sein Schöpfer Dr. Heinrich Hoffmann. Bibliographie der Sekundärliteratur.* Neckarsteinach: Edition Tintenfass, 2015.

Slany, Katrzyna. *Groza w literaturze dziecięcej. Od Grimmów do Gaimana.* Kraków: Wydawnictwo University Press, 2016.

Stahl, John Daniel. "Mark Twain's *Slovenly Peter* in the Context of Twain and German Culture." *The Translation of Children's Literature: A Reader*, edited by G. Lathey. Multilingual Matters, 2006, pp. 211–224.

Szwebs, Weronika. "Tłumacz jako marka? O tym, co zastąpiło kryteria w internetowej recepcji przekładu Wielkiego Gatsby'ego Francisa Scotta Fitzgeralda." *Między Oryginałem a Przekładem*, vol. XXII, 2016, pp. 95–116.

Tahir-Gürçağlar, Şehnaz. "What Texts Don't Tell: The Use of Paratexts in Translation Research." *Crosscultural Transgressions Research Models in Translation Studies II: Historical and Ideological Issues*, edited by T. Hermans, St. Jerome, 2002, pp. 44–60.

Venuti, Lawrence. *The Translator's Invisibility: A History of Translations.* Routledge, 1995.

Wesseling, Elisabeth. "Visual Narrative in the Picture Book: Heinrich Hoffmann's *Der Struwwelpeter.*" *Children's Literature in Education*, vol. 35, no. 4, 2004, pp. 319–345.

Whitehurst, G. J., et al. "Accelerating Language Development Through Picture Book Reading." *Developmental Psychology*, vol. 24, no. 4, 1988, pp. 552–559.

Translating Place and Space: The Soviet Union in North Korean Children's Literature

Dafna Zur

On August 15, 1945, the Korean peninsula was liberated from nearly four decades of Japanese colonial rule. Japan lost the Pacific War and quickly evacuated from the colony it had held since 1910. The last-minute intervention of the Soviet Union in the Pacific War, and the negotiations between the Soviets and the Americans over spheres of influence in this strategic region, resulted in the division of the peninsula along the 38th parallel. This geographical division, intended to be temporary, was cemented through separate military occupations and intense ideological disagreements among Koreans. By 1948, competing governments were established in the north and south, and a destructive civil war erupted in 1950, took millions of lives, and flooded the peninsula with refugees.[1] The Korean War ended three years later with a ceasefire, the precarious status of which continues to this day.

Postwar North and South Korea were faced with the task of recovery, and foreign support proved critical to their abilities to do so. For the Democratic People's Republic of Korea (hereafter, North Korea), recuperation was facilitated by Soviet political, economic, and military presence, particularly in the sphere of education and science (Balázs

D. Zur (✉)
Stanford University, Stanford, CA, USA

A. Kérchy and B. Sundmark (eds.), *Translating and Transmediating Children's Literature*, Critical Approaches to Children's Literature,
https://doi.org/10.1007/978-3-030-52527-9_5

Szalontai 1–34).[2] For example, the new compulsory educational system, established in December 1946, was modeled after the Soviet system, and the Academy of Sciences, first called the *Kwahak ak'ademi* following the Russian appellation, was established in 1952. The budding North Korean science and technology sector was made possible through the aid of Soviet scientists and the training of Koreans in the Soviet Union (Kim Kŭn-bae 376–382). Soviet culture also had a great influence on North Korea, which adopted socialist realism as an aesthetic doctrine, demanding that literature present a "truthful" depiction of life—truth, of course, as defined by the Party (Marshall R. Pihl 77–78).

Among Soviet-inspired cultural forms that received attention was literature for children, which Koreans understood as critical to their political and cultural future. This is apparent in two new genres that appeared in the mid-1950s: travel writing and science fiction. In different ways, these two genres offered a translation of the world in both a literal and figurative sense. Travel writing was punctuated by North Koreans' admiration of Soviet children's culture as a way of calling attention to the work that needed to be done at home; at the same time, they offered opportunities to assert budding North Korean pride. As for science fiction, it was introduced as a new genre through translations from Russian, which brought together scientific education and a socialist moral vision. However, these translations were not imitations of Soviet models. A close reading of translations and their original texts reveal translators' complex negotiation of ideas about gender and science at a critical moment in North Korea's early literary development.

The Soviet Union in North Korean Children's Literature

After the establishment of an independent government in 1948, North Koreans were determined to build a distinct literary culture for young readers. A range of special interest journals for young readers emerged: some were more literary in focus, such as *Adong Munhak* (Children's Literature), *Saesedae* (New Generation), and *Sonyŏndan* (Scouts). Others were more scientific in content, such as *Kwahak Segye* (Science World) and *Sonyŏn Kwahak* (Junior Science). Aimed at a readership ranging from elementary to high-school age, these journals made space for North Korea's writers, poets, educators, and translators to build a vibrant culture for young people that would support the broader social and political nation-building project.

These journals contributed to the building of a uniquely North Korean children's and youth culture, but the shadow of the Soviet Union loomed large. Writers had to negotiate their assertion of a national brand on the one hand, while embracing Soviet influence and support, which was central to North Korean postwar reconstruction, on the other. In a process described as a "wholesale imitation of Soviet Stalinist models" (Gabroussenko 13), writers like Gorky and Pushkin were held as beacons to be emulated, and hundreds of Soviet texts were translated into Korean. This continued until a palpable waning of Soviet influence began in the mid-1950s as part of North Korea's response to Soviet de-Stalinization. Diminished Soviet influence can be traced, for example, in Kim Il Sung's speech at a Korean Workers' Party from 1955, in which he warned against the "neglect of tradition" and complained about the fostering of "a preponderance of all things foreign," citing the prevalence of portraits of Pushkin and Mayakovsky hanging in schools (Gabroussenko 17).[3] Nevertheless, the pervasive presence of the Soviet Union in children and youth culture well into the 1960s puts the argument about waning Soviet influence into question.[4] At the very least, during the decade following the Korean War, Soviet culture served as a model that North Koreans engaged with enthusiastically but also critically, particularly through travel literature and translation.[5]

Travel essays and poetry took their young Korean readers on journeys that stretched from Red Square to the plains of Kazakhstan, from the streets of Moscow and Stalingrad to the personal library of Nikolai Ostrovsky, and from Soviet Pioneer Palaces to trains run by children. But while their geographic imagination was located firmly in the Soviet Union, these writings provide an insight into what was deemed the desirable direction of North Korea's own nascent cultural industry.[6] Three essays in particular—"A Visit to the Zhdanov Palace of Young Pioneers in Leningrad,"[7] "The Children's Train of Stalingrad,"[8] and "Beautiful Moscow"[9]—balance praise for the Soviet Union with expressions of national pride. In Stacy Burton's words, "travel literature reveals much about the construction of the self, the representation of experience, the ideologies of colonialism and imperialism, the boundaries between fact and fiction, and the relations between readers and texts" (51). In this sense, North Korean travel literature is as much designed to document the places traveled as to pointing to the direction toward which North Korean children's culture should strive to develop.

Kang Hyo Sun's "A Visit to the Zhdanov Palace of Young Pioneers" opens with praises of Leningrad. In the narrator's words, this "historic" and "heroic" city bubbles with stories about its peoples' brave resistance to the German siege in the Second World War. The Zhdanov Palace of Young Pioneers[10] stands out: it is "magnificent" and "beautiful," wrested from the "exploitative" hands of the Czars by the Communist Party during the October Revolution and "bestowed, in its infinite grace, upon Soviet children for the sake of their happiness and future" (Kang, "Chŭdanobŭ" 40–41). The discerning eye of the narrator attempts to capture every visual detail and convey it to the readers back home. His attention is drawn by two encounters, one literary and the other scientific/technical. In the "Writers' Room," Kang meets a fifteen-year-old Russian girl reading a book of translated Korean folk tales that contain the "life and feeling" of the Korean people. She delights to learn that the visitor is Korean, and promises to continue reading Korean books. That a Korean book is the one she is reading among the professed 104,000 volumes in the library's collection might serve to stoke pride in the hearts of the readers back home. In addition, the narrator's detailed description of the reading rooms decorated with statues and illustrations of famous Russian writers and stories contains an implicit message to value and cultivate love for Korea's own tradition.

Another impressionable encounter takes place as the writer wanders through rooms where vocational and technical knowledge is imparted in a most entertaining manner. He observes, "All the facilities seem to be designed for entertainment purposes…but they set children up, in a natural way, to conduct scientific experiments." Literature and science come together in a wood-crafting room where children make wooden models of scenes from their favorite tales. "Seeing the happiness of the children raised under the Socialist system fills me with content, for I know that our own children of the North Korean Republic are already well on their way toward this same kind of happiness" (Kang, "Chŭdanobŭ" 43–45). This aspirational conclusion reminds young readers that they are part of a larger republic of socialist children working with their hearts, minds, and hands toward building a common utopian future. Kang's "The Children's Train of Stalingrad" has a similar structure. After taking stock of the grandeur of Stalingrad, a city whose war scars are still palpable, Kang observes that "Everywhere we turn, the boundless generosity that the

Soviet Union bestows upon its children, in the form of children's establishments, greets us with great rapture, and this was no different with the children's train" ("Ssŭttaringŭradŭ" 45). What impresses Kang is the respect and love that the Soviets have for their children—and implied here is that North Korea should adopt a similar attitude. If the essay about the Zhdanov Palace focused mostly on the literary encounter, this essay celebrates the Soviets' achievements in the scientific and technical, which is represented by the child-operated train. The experience allows Kang to marvel at the combination of theory and practice that is exemplified by this feat of engineering, with all the implied metaphors of modernity, speed, and connectivity. Among this celebrated achievement, which the writer can only hope to replicate in Pyongyang, he still finds a moment to assert North Korean pride. When the passengers learn that there are Koreans on the train, they rush over to meet them and request that their names be written in Korean. He reflects on the affection with which he and his team are received, and how proud he is to be traveling in the USSR as a "victorious North Korean" (50).

In "Beautiful Moscow," Ri Wŏn U records his visit to Moscow, which he describes as part of the "nation of the happiest lives and the creators of the most beautiful civilization" (6). Like Kang, Ri meditates on the city's recent history, particularly how the statues of Lenin seem to be directing the children of the Soviet Union forward with such charisma that Ri can almost hear Lenin's voice (8). All that Ri observes, from the department store to the gardens of Moscow, speaks of the Russian attention to the needs of children and is contrasted with America's neglect of its own. Russians sell educational toys that develop children's knowledge of science; Americans sell their children guns. Russian department stores are filled with customers; American children languish in poverty. Ri observes, "The items on display in Soviet department store are not intended for making profit. *They have been created to make beautiful and lofty the cultural lives of children*" (12; emphasis in original). Ri makes these observations as a way of encouraging the development of children's culture back home. The opportunity to assert his North Korean identity comes at the chance meeting with seventy-year-old Samuil Marshak. Accompanied by another North Korean literary luminary, Pak Se Yŏng, the two introduce themselves and Ri advises the Russian luminary, "Please take care of yourself. Your body is valuable not only for Soviet children, but to children from all over the world." To this, Marshak responds, "You seem to have a weak disposition yourself; for the sake of the children let us not get old, let us not get sick, let us struggle to write good work!" (16–17). By quoting Marshak, Ri seems to be paying himself a compliment by placing himself on the same level of importance, acknowledging their shared concerns over health and their literary mission.

Tatiana Gabroussenko argues that North Korean travel accounts from this period were "deliberate fantasies" full of gross misrepresentation and straight out "untruths" performed by North Korean delegates, either as acts of loyalty to the North Korean regime or out of a propensity for sentimentalism (28–45). However, whether travel accounts were deliberately misleading or products of carefully curated tours by the Soviet hosts is beside the point. It is not truth or authenticity that should be accounted for in travel literature since the role of the travel writer is closer to that of the novelist (Adams 280–281): North Korean writers, asserting their subjectivities, sought to create an experience of "being there" that was meant to be evocative and stir developments at home. The act of travel, as James Clifford argues, is one that evokes a complex range of experiences that trouble "the localism of many common assumptions about culture." Clifford shows that travel, when acting as a supplement to cultural life, might emerge "as *constitutive* of cultural meanings" (3). Clifford finds that the process of locating oneself in time and space is one that involves a series of "encounters and translations" (11); this means establishing concepts built on "imperfect equivalences." I argue that travel instantiates the process of what I call "translating space." In the process of conveying the "text" of the cities they toured, Korean children's writers strove to recreate an experience colored by their own aspirations, and the stories they chose to tell had the dual function of reaffirming North Korean values through the Soviet models before them.

The Future in Translation: Soviet Science Fiction in North Korea[II]

From the start, children's literature played a critical role in North Korea's postwar transformation. In the words of critic Kim Myŏng Su, "The task of proper cultivation of our children is critical to the creation of a prosperous future as well as for our happiness and freedom. Children's literature is that national industry that is charged with bearing fruit of the glorious responsibility" (97). In the same breath, Kim Myŏng Su evoked Tolstoy, Gorky, and Marshak as the models to which writers should aspire (98–99). These writers were celebrated for the way they brought together politics and aesthetics that both described and prescribed the socialist experience. From its inception, the journal *Adong Munhak* included translations in each issue, including works by A. Aleksin, I. Krylov, N. Bogdanov, G. Leonidze, A. Kononov, A. Gaĭdar, and many others. Then, in April 1956, B. Li͡apunov's "We Landed on Mars" appeared in Korean translation.[12] It marked the advent of the new genre of science fiction that straddled realism and fantasy, a genre spurred by North Korea's excitement over science and technology and shaped by socialist values of control over nature and collective striving.

Li͡apunov's was the last science-fiction translation to appear in *Adong Munhak*; thereafter, Korean writers published their own original pieces.[13] Translations of Russian science fiction and nonfiction continued to be published in another venue: *Saesedae* (New Generation), a literary journal that targeted middle- and high-school readers. Among the writers translated in *Saesedae* were V. Shevchenko, G. Gurevich, A. Torohov, Iu. Safronov, V. Morozov, and N. Erdman. In most cases, the translations were fairly close to the original, diverging from the original only with minor adjustments in content and style. For example, in the translation of "Secrets of the Deep Sea,"[14] the translator writes that "the seas surrounding Japan are rife with sharks,"[15] when the original word was "sardines," mistranslated perhaps for greater dramatic effect. Some discrepancies between the original and translations can be attributed to misreadings. For example, in the translation of "In Year 2017,"[16] the translator misread "Alpha Centauri" as the name of the rocket instead of the name of a galaxy.[17] In some cases, the translators opted for flowery language not present in the original. The translator of "Nothing Left of Enchantment"[18] inserted metaphors bordering on the cliché, such as "the water's surface was like a mirror," "cold sweat ran down his back," or "he didn't know if he was dreaming," none of which were present in the original. These clichés can be viewed as places where the translator's voice asserted itself, reproducing what Nida et al. might call a "dynamic equivalent" (ch. 2). Against the assertion that translations offer proof of wholesale imitation of the Soviet original, such metaphors stand out as moments where local flavor makes itself apparent.

Of the Russian works published in *Saesedae*, the translation of the animated script "Flight to the Moon"[19] has the most interesting and revealing gaps between the original its translation. The Korean text, published in 1956, is a translation of the thirty-minute animated film from 1953, not the full-length Russian script that was published in 1955.

What sets this translation apart from the others discussed above is that the Korean translators took great liberties by redacting, supplementing, and in some cases, changing the content. The most notable of these discrepancies appear in relation to the female character Natasha. Natasha is the only girl in the story and is not a member of the three-member crew "International Society for Interplanetary Communications named after Tsiolkovsky" consisting of Kolya the Russian, Petya the Ukrainian, and Sandy the African-American. However, Natasha becomes an important character because her father, a rocket scientist whom the boys admire, has just set off to the moon, and the "International Society" members observe his mission from their rooftop telescope. When Kolya and his friends learn that the rocket has been stranded on the moon, their first thought is of their comrade Natasha. Natasha is heartbroken about the news, and she asks the boys to help send her clever dog, Toby (renamed "Padugi" in the Korean), to save her father. By comic accident, Kolya ends up traveling to the moon as a stowaway in the crate meant for the dog. There, his nimble actions bring about the safe return of Natasha's father.

In similar fashion to other translations discussed above, the translators of this animated film took liberties in their translations, editing some sections and adding others. For example, the opening of the Russian film has Sandy listening eagerly for news of the mission to the moon while Kolya searches for the rocket through his telescope. However, the Korean translation opens with an argument between Kolya and Petya about whether the moon has mountains, water, and beasts.[20] The narrator of the Korean translation also inserts a commentary, missing from the Russian, that Kolya's dream is to become an astronomer like Natasha's father. From the very opening, then, the Korean narrative diverges from the Russian film in that Natasha's father is the subject of emulation, not space travel more generally.

The discrepancies between the original Russian and the Korean version are due in part to the shift in genres—the animated film has wordless scenes where the action is on the screen, accompanied by music that ranges from sinister to triumphant, making verbal narration redundant. It is in these places that the Korean translators sought to employ onomatopoeia and metaphors to produce rich descriptions of scenery. The most interesting divergence from the original takes place when Petya and Sandy leave the takeoff site with what they think is a crate containing their friend Kolya. In the original film, they open the crate and discover that it is empty; their attention is then drawn to the bright lights of the rescue-ship takeoff. Only after that do they discover that they had the wrong crate. In the Korean version, the two boys do not find the box with their friend, and when they see a streak of light across the sky, they vent their frustrations on Natasha and on the dog, less because they lost track of Kolya and more because they missed the rocket launch. For her part, Natasha, in the Korean version, feels bad because she failed to send her dog to save her father.[21] In the Russian version, Natasha is a vulnerable character the boys feel committed to protecting, whereas in the Korean she is portrayed as foolish and immature, and therefore deserving of the boys' ridicule. Such discrepancies call attention to the extent to which the Korean translators were working through gendered conceptions that did not result in a "faithful" translation.

Riitta Oittinen interrogates the concept of equivalence, which is the degree to which a translation is the "same" as the original. Equivalence, as she points out in her survey of translation theory, remains in the realm of the ideal. Translators bring to the table ideologies and norms that emerge from their own interpretive communities. She draws attention to the imagined audience facing translators of children's literature, because it is translators' ideas about their target audience that drive their decisions. In other words, the imagined audience in literature for children plays a bigger role in shaping translations than is the case for adult literature (Oittinen 76). In Oittinen's words, translations do not "produce sameness" but bring to the fore a range of interpretive horizons and interpretive communities because the understanding of what child culture is can vary drastically from one location to the next (161). Gillian Lathey also argues that translators writing for a child readership "work in real 'geopolitical situations' that determine translation practice," and they either conform to or challenge constructions of childhood precisely because of the volatility of that concept (6). Translators for children, Lathey notes, "are mediators not just of unfamiliar social and cultural contexts, but also of the values and expectations of childhood encoded in the source text" (196). In holding up travel literature against "real" cities and translations of Soviet fiction against their original texts, the moments where North Korean writers assert their own interpretations are the most revealing. They do not tell the story of how accurately the translators read Russian or how eagerly they followed the Soviet propaganda machine. Rather, the discrepancies reveal the ideological and future aspirations of a nation truly invested in building a scientific, technical, and socialist utopia.

Notes

1. For readings on the Korean War in English, see Bruce Cumings, *Korea's Place in the Sun*, W.W. Norton, 2005; Bruce Cumings, *The Korean War*, Modern Library, 2011. In addition, *North Korea Caught in Time* provides an introductory essay by Szalontai accompanied by photographs of wartime North Korea. For articles on the Korean War in children's literature of North and South Korea, see Dafna Zur, "Representations of the Korean War in North and South Korean Children's Literature," in *Korea 2010: Politics, Economy, Society*, edited by Rüdiger Frank, James E. Hoare, Patrick Köllner, and Susan Pares, Brill, 2011, pp. 271–300; Dafna Zur, "'Whose War Were We Fighting?' Constructing Memory and Managing Trauma in South Korean Children's Fiction," *International Research in Children's Literature* 2.2, 2009, pp. 192–209; and Dafna Zur, "The Korean War in Children's Picturebooks of the DPRK," in *Exploring North Korean Arts*, edited by Rüdiger Frank, Nuremberg: Verlag für moderne Kunst, 2001, pp. 276–298.
2. For English language scholarship on the early history of North Korea, see A. N. Lankov, *From Stalin to Kim Il Sung: The Formation of North Korea, 1945–1960*, Hurst, 2002; Charles Armstrong, *The North Korean Revolution*, Cornell University Press, 2003; Bruce Cumings, *North Korea*, New Press, 2004; Balázs Szalontai, *Kim Il Sung in the Khrushchev Era*, Woodrow Wilson Center Press, 2005; A. N. Lankov, *Crisis in North Korea: The Failure of de-Stalinization, 1956*, University of Hawai'i Press, 2005; and Suzy Kim, *Everyday Life in the North Korean Revolution*, Cornell University Press, 2013.
3. See also Hyung-chan Kim and Tong-gyu Kim, *Human Remolding in North Korea: A Social History of Education*, University Press of America, 2005, pp. 76–77.
4. To be fair, Gabroussenko acknowledges that "a large number of stereotypes that had entered Korea in the late 1940s from the USSR survived this de-Russification and have remained an important part of the North Korean literary tradition to this day." But I believe that it is too facile to dismiss the presence of Soviet influence in North Korea in the period after the mid-1950s as "stereotypes."
5. While this essay focuses on postwar North Korea, both prerevolutionary Russian culture and post-revolution Soviet culture had a lasting influence on the Korean peninsula long before division. See Heekyoung Cho, *Translation's Forgotten History: Russian Literature, Japanese Mediation, and the Formation of Modern Korean Literature*, Harvard University Asia Center, 2016; Vladimir Tikhonov, "Images of Russia and the Soviet Union in Modern Korea, 1880s–1930s: An Overview," *Seoul Journal of Korean Studies*, vol. 22, no. 2, 2009, pp. 215–247.

6. *Adong munhak* [Children's Literature] was first issued in 1946, and it has been published in monthly installments since January 1954. It was the representative publication of the North Korean writers' organization, the *Chosŏn chakka tongmaeng chungang wiwŏnhoe* (North Korean Writers' Federation), and saw as its mission the education and cultivation of North Korean children. See Wŏn Chongch'an, 2012, pp. 238–243.
7. Kang Hyo Sun, "Chŭdanobŭ ppionerŭ kungjŏn." Kang (1915–1983) was one of North Korea's most prolific children's writers.
8. Kang Hyo Sun, "Ssŭttaringŭradŭ adong ch'ŏlto."
9. Ri Wŏn U, "Arŭmdaun mossŭk'ŭba." Ri (1914–1985) was a celebrated writer and critic of literature for children who was active throughout the colonial period; he penned one of North Korea's classic works of fiction of resistance titled "General Ax."
10. The Zhdanov pioneer palace was converted from the Anichkov Palace in 1934 and became one of the most famous palace of young pioneers in the Soviet Union. Pioneer palaces offered classes in politics, technical skills, art, sports, tourism, agriculture, reading, and education (*Adong munhak* 1954, pp. 8, 41). On the Moscow Pioneer Palace, see Susan Emily Reid, "Khrushchev's Children's Paradise: The Pioneer Palace, Moscow, 1958–1962," *Socialist Spaces: Sites of Everyday Life in the Eastern Bloc*, edited by David Crowley and Susan Emily Reid, Berg Publishers, 2002, p. 141.
11. My profound gratitude goes to Byungsam Jeong, Ph.D. candidate in Stanford University's Slavic department, who read the Russian originals and pointed out the discrepancies between the Russian and the Korean translations.
12. Song Tong Kyu, translator. "Uri dŭl ŭn hwasŏng e watta," *Adong munhak*, April 1956.
13. For a discussion of North Korean science fiction, see Dafna Zur, "Let's Go to the Moon," 2014; and Dafna Zur, *Figuring Korean Futures*, 2017.
14. "Pada mit ŭi pimil," *Saesedae*, May–June 1955. It was originally titled "Taĭny podvodnogo mira" by A. Dorohov.
15. Ibid., June 1955, p. 32.
16. "30 nyŏn twi e," *Saesedae*, January 1958, pp. 41–47. It was originally titled "V 2017 godu" by V. Shevchenko.
17. Ibid., p. 42.
18. "T'ŭkbyŏlhan kŏsŭn ŏpta," *Saesedae*, May–July 1960. It was originally titled "Nichego Osobennogo" by Iu. Safronov.
19. See "Tallara chajasŏ," *Saesedae*, March 1956, pp. 52–54; "Tallara chajasŏ," *Saesedae*, April 1956, pp. 41–50. It was originally titled "Polyot na Lunu" by V. Morozov and N. Erdman. The original Soviet version with English subtitles can be viewed here: https://www.youtube.com/watch?v=wrv4X80xGaY.
20. Ibid., March 1956, p. 52.
21. Ibid., April 1956, pp. 42–43.

Works Cited

Adams, Percy G. *Travel Literature and the Evolution of the Novel*. The University Press of Kentucky, 1983.

Armstrong, Charles K. *The North Korean Revolution, 1945–1950*. Cornell University Press, 2003.

Burton, Stacy. "Experience and the Genres of Travel Writing: Bakhtin and Butor." *Romance Studies*, vol. 11, no. 1, 2013, pp. 51–62.

Chang, Mi-Kyoung. *A Critical Content Analysis of Korean-to-English and English-to-Korean Translated Picture Books*. Dissertation, University of Arizona, 2013.

Cho, Heekyoung. *Translation's Forgotten History: Russian Literature, Japanese Mediation, and the Formation of Modern Korean Literature*. Harvard University Asia Center, 2016.

Clifford, James. *Routes: Travel and Translation in the Late Twentieth Century*. Harvard University Press, 1999.

Cumings, Bruce. *North Korea: Another Country*. New Press, 2003.

———. *Korea's Place in the Sun*. W.W. Norton, 2005.

David-West, Alzo. "Marxism, Stalinism, and the Juche Speech of 1955: On the Theoretical De-Stalinization of North Korea." *The Review of Korean Studies*, vol. 10, no. 3, 2007, pp. 127–152.

Frank, Rüdiger. *Exploring North Korean Arts*. Nürnberg: Verlag für moderne Kunst, 2011.

Gabroussenko, Tatiana. *Soldiers on the Cultural Front: Developments in the Early History of North Korean Literature and Literary Policy*. University of Hawai'i Press, 2010.

Kang Hyo Sun. "Chŭdanobŭ ppionerŭ kungjŏn" ["A Visit to the Zhdanov Palace of Young Pioneers"]. *Adong munhak*, August 1954, pp. 40–45.

———. "Ssŭttaringŭradŭ adong ch'ŏlto" ["The Children's Train of Stalingrad"]. *Adong munhak*, October 1954, pp. 44–50.

Kim, Hyung-chan and Kim, Tong-gyu. *Human Remolding in North Korea: A Social History of Education*. University Press of America, 2005.

Kim, Kŭn-bae. "'Pukhan kwahak kisul ŭi yŏksajŏk chŏngae.'" *Pukhan ŭi kyoyuk kwa kwahak kisul*, edited by Pukhan yŏn'gu hakhoe, kyŏngin munhwasa, 2006, pp. 375–400.

Kim Myŏng Su. "Adong munhak ch'angjak e issŏsŏ ŭi myŏkkaji munje." *Chosŏn munhak*, December 1953, pp. 97–116.

Kim, Suzy. *Everyday Life in the North Korean Revolution, 1945–1950*. Cornell University Press, 2013.

Lankov, A. N. *From Stalin to Kim Il Song: The Formation of North Korea, 1945–1960*. Hurst, 2002.

———. *Crisis in North Korea: The Failure of de-Stalinization, 1956*. University of Hawai'i Press, 2005.

Lathey, Gillian, editor. *The Translation of Children's Literature: A Reader*. Multilingual Matters, 2006.

———. *The Role of Translators in Children's Literature: Invisible Storytellers*. Routledge, 2010.

Nida, Eugene A., Taber, Charles, and Brill, E. J. *The Theory and Practice of Translation*. Brill, 1969.

Oittinen, Riitta. *Translating for Children*. Garland, 2000.

Pihl, Marshall R. "Engineers of the Human Soul: North Korean Literature Today." *Korean Studies*, vol. 1, no. 1, March 2011, pp. 63–92.

Reid, Susan Emily. "Khrushchev's Children's Paradise: The Pioneer Palace, Moscow, 1958–1962." *Socialist Spaces: Sites of Everyday Life in the Eastern Bloc*, edited by David Crowley and Susan Emily Reid, Berg Publishers, 2002, pp. 141–180.

Ri Wŏn U. "Arŭmdaun mossŭk'ŭba," *Adong munhak*, October 1957, pp. 4–17.

Springer, Chris. *North Korea Caught in Time: Images of War and Reconstruction*. Garnet Pub., 2010.

Szalontai, Balázs. *Kim Il Sung in the Khrushchev Era: Soviet-DPRK Relations and the Roots of North Korean Despotism, 1953–1964*. Woodrow Wilson Center Press, 2005.

Tikhonov, Vladimir. "Images of Russia and the Soviet Union in Modern Korea, 1880s–1930s: An Overview." *Seoul Journal of Korean Studies*, vol. 22, no. 2, 2009, pp. 215–147.

Wŏn Chongch'an. *Pukhan ŭi adong munhak: Chuch'e munhak e irŭnŭn tojŏng*. Kyŏnggi-do P'aju-si: Ch'ŏngdong kŏul, 2012, pp. 238–43.

Yi, Hyanggyu. "Pukhan sahoejuŭi kyoyuk ŭi hyŏngsŏng: kyoyuk kihoe hwakdae rŭl chungsim ŭro." *Pukhan kwahak kisul hyŏngsŏngsa*, edited by Kang Hoje, Sŏnin, 2007, pp. 7–43.

Zur, Dafna. "Let's Go to the Moon: Science Fiction in the North Korean Children's Magazine Adong Munhak, 1956–1965." *The Journal of Asian Studies*, vol. 73, no. 2, 2014, pp. 327–351.

———. *Figuring Korean Futures: Children's Literature in Modern Korea*. Stanford, 2017.

Image-Textual Interactions

"How Farflung Is Your Fokloire?": Foreignizing Domestications and Drawing Bridges in James Joyce's *The Cat and the Devil* and Its French Illustrations

Aneesh Barai

The Cat and the Devil, a story James Joyce wrote in 1936, originally as a letter to his grandson Stephen, became a children's classic once it was made into a picturebook in 1964. Although the story is predominantly written in English, it also contains French and Italian elements: the devil speaks French to the townspeople for a page (fifty-one words), and the original letter Joyce wrote is signed off as "Nonno," which is Italian for "grandfather." The cumulative effect of these language choices is a polylingualism expressive of the international nature of modernism. Although *The Cat and the Devil* (hereafter *Cat*) has been translated into thirteen languages from the 1960s onwards, only one article has been written on issues relating to its translation.[1] Only two years after the publication in England, the first French translation appeared as *Le Chat et le Diable* (hereafter *Chat*) by Jacques Borel, who went on to translate Joyce's *Chamber Music* and *Pomes Penyeach*. It has had two French illustrators: Jean-Jacques Corre and Roger Blachon.

A. Barai (✉)
University of Sheffield, Sheffield, UK

A. Kérchy and B. Sundmark (eds.), *Translating and Transmediating Children's Literature*, Critical Approaches to Children's Literature,
https://doi.org/10.1007/978-3-030-52527-9_6

The translation of *Cat* into French complicates issues of domestication and foreignization (as I will define below) and strains the relationship of language to place: *Cat* tells the local legend of Beaugency, a town that Joyce visited in France. In this tale, the devil builds a bridge for the town and is cheated of his payment of the first person to cross the bridge when the mayor makes a cat cross it first. As a local legend, its interest in location is clear and reinforced by Joyce's choice of Beaugency's version of the legend over variants that he would have read in Henry Bett's *Nursery Rhymes and Tales*. This folk tale exists in different forms around the world, and it is in the Aarne-Thompson standard directory of folk tale motifs as motif number 1191. It is essential to see that Joyce's story is thus also a translation, of a tale he heard in Beaugency, and as such, we can see his approach to the story as commensurate to the work of a translator. Joyce makes key choices in adapting details of the story in order to best engage his target audience, his grandson.

Lawrence Venuti's seminal work in translation studies sets out the central debate between domestication and foreignization. Venuti distinguishes between translations that choose to domesticate foreign details and those that choose to foreignize—that is, to keep foreign details foreign—and determines that this is a fundamental divide in attitudes toward translation (*The Translator's Invisibility*, *passim*). With these terms, Venuti gives names to the decision process that Friedrich Schleiermacher outlined as early as 1813: "Either the translator leaves the author in peace as much as possible and moves the reader towards him; or he leaves the reader in peace as much as possible and moves the author towards him" (cited in Venuti, *The Translator's Invisibility* 15). Furthermore, the particular challenge of translating picturebooks must take into account interactions between words and images and the impact of illustrations. In *Comparative Children's Literature*, Emer O'Sullivan points to potential pitfalls when translating picturebooks. For example, she writes of a German translation of Philippe Dumas' picturebook *Laura sur la route* (1978), which domesticates all the place names but still has the Eiffel Tower in its pictures (85). She notes that some publishers provide directives to avoid such mistakes by enforcing a "pre-censorship" of culturally specific material, encouraging what she calls an "international insipidity" (86–87). O'Sullivan points to the damaging consequences of these efforts by publishers to remove culturally specific items from text or image:

> Instead of multiculturality based on knowledge and acceptance of the differences between cultures, we have here an (alleged) cultural neutrality, resulting in non-specific, levelled-out, international products. The mere fact that children's literature is being translated or coproduced thus has no particular cultural value in itself. (88)

Beyond this negative process of reducing images to cultural neutrality, there is nothing written on the potential positive impact of commissioning new target-culture illustrators for translations, as is the case for Joyce's *Cat* in France. *Cat* is a particularly rare example, as the French illustrator Blachon's work was so popular that his illustrations are now also used for English-language editions of the text. In the same way that translators face decisions of domestication, foreignization, and modernization and hold a particular notion of childhood in mind for which they tailor their work, native illustrators of translated texts bring significant notions of culture to their illustrations. This chapter argues that the two French illustrators of Joyce's text push against the foreignization that takes place in its verbal translation by drawing on French culture in their images.

I will first consider the translation decisions that Joyce made in reworking this French local legend, then consider how French translators have sought to accommodate Joyce's language in their texts, and finally address the fraught encounter that troubles the binary distinction of "domestication" and "foreignization" that occurs when back translation is an issue in a story that thematizes language. Joyce overlaps cats and letters in their textual functioning as bridges between cultures and, at the same time, domesticates this French local legend, as the devil and the mayor take on contemporary Irish traits. Joyce's French translators not only leave these Irish elements as strained foreignizations but also seek to repeat their effect—by foreignizing what was originally domestic: the French language itself. I will finally consider the role of illustrations in translated children's literature and look at how the French illustrators Corre and Blachon successfully find a middle ground in cultural representations—not only illustrating the bridge at the heart of the story but also building bridges between cultures through their images, connecting the picturebook with modernist art styles (for Corre), and forging a French-Irish Catholic link (for Blachon). Riitta Oittinen considers the relation between image and text through the Bakhtinian lens of dialogism (*Translating for Children* 101), and this provides a useful tool to consider the polyvocality of these translations of Joyce's picturebook, with text and image voicing different cultures and thus the text as a whole (the iconotext) disrupting the binary distinction of domestication and foreignization. Illustrations, I argue, have the potential in picturebook translations to sit between domestication and foreignization as categories, through the commonalities that they draw between cultures.

Joyce's Translation from France to Ireland

Where *Finnegans Wake* asks, "How farflung is your fokloire?" (419.11–12), evoking a particular link between folklore and the river Loire, *Cat*'s answer appears to be "close to home" in domesticating differences, as Joyce translates Beaugency's legend for his grandson.[2] Joyce translates this legend into a modernist text with an Irish focus, which is of his own style, but for a child. We can see Joyce's efforts to bridge between the story world and his target audience's knowledge in his adaptations, for example, explaining in the story that the Loire is "France's longest river." While the tale is originally one of rebelling against authority and outwitting the devil, Joyce makes it his own by reversing this and siding with the devil as the outwitted party; however, he suggests some consolation in this position, as the exiled devil gains the company of a scapegoated cat.

What is most notable in Beaugency's version of the tale is that the creature made to cross the bridge is a cat, from which the people of Beaugency have earned their (still extant) nickname, "les chats de Beaugency" ("the cats of Beaugency"). In Bett's versions, it can be a goat, a rooster, a dog, or a cat that crosses. Patricia Dale-Green, the catlore specialist, writes of the Beaugency myth, "Satan was so furious when he found he had been fobbed off with a cat that he tried to kick down the new bridge. He failed, however, and as he carried off the cat, it tore at his hands and face with its claws" (131–132). Joyce's cat and devil make a somewhat friendlier team, however, with the devil affectionately calling the cat "mon petit chou-chat" ("my little sweetheart-cat") and taking it off to dry it. Joyce himself was particularly fond of cats, and Frank Budgen tells us that he owned a black cat (321). Cats and mayors come together elsewhere in Joyce, as Dick Whittington turns up in *Wake*. The mayor in *Cat*, Alfred Byrne, also appears in a mocking list in *Wake*, in the Dom King episode. He is not mayor here, but the mayor Pomkey Dompkey reads, "His Serenemost by a speechreading from his miniated vellum, *alfi byrni* gamman dealter etcera zezera" (568.31–32; my italics). Byrne is made diminutive by the alphabetizing and by the unstressed rhymed endings to his inclusion. *Cat* shows that such an insulting allusion is part of a trend in Joyce's interpolation of Byrne.

Bridges hold a central position in Joyce's story. While Hope Howell Hodgkins sees the bridge as a metaphor for the relation of adult author to child reader (363), it can also be seen as a powerful metaphor for the cross-cultural narrative of Joyce's letter and the cross-cultural performance of translating it. The letter itself can be seen as a bridge across countries, as Joyce relates his travels around Europe to his grandson. All editions of *Cat* not only maintain the letter format but most also keep the dateline, with the location in it, to indicate that the story comes from abroad. Further, Jacques Charpentreau, the French translator of T. S. Eliot's *Old Possum's Book of Practical Cats*, sees cats as bridges between cultures since children anywhere in the world can relate to them: "Les chats [forment] une aristocratie internationale [...] Car nous avons aussi nos chats pirates comme Grostigré (ils ne hantent pas la Tamise, mais la Seine)" ("Cats form an international aristocracy... For we also have our pirate cats like Growltiger [they do not haunt the Thames, but the Seine]"; *Chats!* 6). Moreover, cats are traditionally seen as bridges between worlds. As Dale-Green describes, "The cat formed a bridge not only between good and evil, but also between interior and exterior life, and between supernatural forces and men" (132).

Joyce's letters and postcards often talk of cats to Stephen, at times acting as supplements and substitutes for cats in themselves, just as the cat acts as a substitute for a human soul in the legend. For example, in a later letter to Stephen from Copenhagen (now published as *The Cats of Copenhagen*), Joyce apologizes that there are no cats there to describe and proceeds to comically depict the city for his grandson. In this way, we can see both that cats are a means of linking the new locations of Joyce's travel to something that a four-year-old will be familiar with, and that the letter serves as a cat substitute in replacing a cat, or a cat-related souvenir, with a playful depiction of a city. The substituting of a cat with a letter is implied even earlier between Stephen and Joyce in a postcard that Stephen received for his second birthday, on which there is a picture of Puss-in-Boots.

In reverse to the movement of texts in place of cats, Marie-Dominique Garnier explores the cat in *Ulysses* as a textual space: "The cat's erect tail, glossy hide and lithe black form call forth the pen, the vellum, the ink and the glossary—a smooth writing space" (104). Lucia Joyce also received a letter about a cat around the mid-1930s, in which Joyce wrote in Italian, "Il famoso gatto, diventato randagio, fa il giro di tutti gli appartamenti e si fa mantenere da tutti con un cinisimo quasi nobile" ("The famous cat which went astray is now making a tour of all the flats and lives on everybody with an almost noble cynicism"; *Letters of James Joyce III* 355). In this way, Joyce's cat letters bring together multilingualism, his children, and mobility. Cat-letters can thus be seen as apt vessels for the cross-cultural and mobile schematics of modernism and, moreover, matched specifically for children.

In *Cat*, it is clear that the story is not a substitute but a supplement to a cat souvenir: "I sent you a little cat filled with sweets a few days ago but perhaps you do not know the story about the cat of Beaugency." Gerald Rose's illustrations interpolate not only the sweet-filled cat but also the letter itself, adjacent to it, and the scene of writing preceding it, with Joyce at his desk. Blachon's begins with the postman delivering the cat, to hint at the commensurateness and codependence of cat toy and story. With both the letter as taking the place of cats and cats as in themselves textual, *The Cat and the Devil* literally writes *on* (about) a cat, telling the story behind a cat toy that Joyce had sent to Stephen, using the full implications of cat and letter to bridge the space between grandfather and grandson.

Among the many playful and anachronistic details that Joyce brings to the tale to engage Stephen, the most noticeable is the naming of the mayor of Beaugency "Monsieur Alfred Byrne," whose French title firmly points to how out of place his name is. Alfred Byrne was the mayor of Dublin in 1936, for the seventh year running. It also happens that Byrne and Joyce were the same age, a fact that may have tinged his hatred of the mayor and his comparisons of himself and Byrne. Joyce particularly hated Byrne for his love of pomp and ceremony. For example, in a letter to Stephen's parents, Joyce tells that he has been invited to the USA, and after mocking "every old fool in Europe" who is invited and goes, Joyce writes, "I see the Lord Mayor of Dublin Alfie Byrne is going to N[ew]Y[ork] for the 17th. Every day I open the *Irish Times* I see him and his golden chain in some photograph or other" (*Letters III* 345–346). Both his gold chain and the scarlet robe that Byrne loved to wear as signs of his office appear in *The Cat and the Devil*, in which he is first gently mocked for his love of pomp and then made wholly ridiculous for his strange habits: "This lord mayor was very fond of dressing himself too. He wore a scarlet robe and always had a great golden chain round his neck even when he was fast asleep in bed with his knees in his mouth." Byrne is shown as a lover of spectacle—announcing his arrival with fanfare—which the cat playfully undermines through its disinterest: "he was tired of looking at the lord mayor (because even a cat gets tired of looking at a lord mayor)." This page further undermines Byrne by stating his title five times in two sentences, such that "lord mayor" becomes meaningless through repetition.

Joyce domesticates the devil, too, in line with the domestication of his parallel, the mayor. For example, Garnier notes that both the devil and the mayor are "compulsive dressers" (100). In Blachon's illustrations, they are both dressed in red robes, have large noses, and as they shake hands on pages 11 to 12, appear to mirror each other. The devil is turned from the epitome of otherness, as he is in all other versions of this folk tale, into a humane, Joycean artist. He is also brought out of sync with history through his anachronistic reading of newspapers and the use of a spyglass. As such, he becomes identifiable with the present and, thus, aligned with the reader. Most illustrators fashion a Joycean devil, complete with goatee and spectacles. Within the text, Rose first positions this Joyce-like figure framed by a mirror, which Amanda Sigler suggests highlights representation and recognition, to evoke that this is not the devil himself but the author (541). Rose also begins with a picture of Joyce writing the letter, hunched over his desk. While we cannot see his face and he has no horns, he looks enough like the devil on the front cover (and five pages later) that one cannot help but identify them as the same figure. Corre's illustrations make this explicit, as he places Joyce's face mid-line after the first mention of the "Diable" ("Devil") and uses the same face on the front cover of the book, just beside the word "Joyce." The author at times associated himself with the devil, acting the devil in family plays as a child and being called "Herr Satan" by his Zurich landlady (Sigler 542). In this way, we may see the model of the author-god turned on its head, bringing to bear the figure of the author-devil. Where the author-god suggests authority and power, the model of the author-devil suggests equally superhuman creative power, yet tied to a figure of exile, of no authority, who gains no respect or reward for his actions. The text itself makes clear the association between the devil and Joyce in its well-known postscript: "P.S. The devil mostly speaks a language of his own called Bellsybabble which he makes up himself as he goes along but when he is very angry he can speak quite bad French very well though some who have heard him say that he has a strong Dublin accent."[3] The devil is not presented as a hateful character: the contemporary and domestic image of the devil in *Cat* makes him a sympathetic figure, and clearly one of more compassion than the mayor.

Sigler draws on these domestications and adaptations of the story to assert that the Balgentiens (people of Beaugency) are aligned with the Irish, and as such, the Irish are here criticized as cowardly and cunning for sacrificing another to cheat a sympathetic devil of his dues. She cites Joyce's declaration in his letters that the Irish people need "one good look at themselves in my nicely polished looking-glass," which Joyce is not explicitly saying about this story but which can certainly fit its domesticating strategy, too (545).

Translating French and Neologisms into French

We may think of translating Joyce's story into French as, in a sense, "untranslating" it back into French. However, in using a French village as a mirror to Dublin, *Cat* plays with the relation of place and language in challenging ways. Joyce's domestication of the mayor and the devil reverses the understanding of events from his source tale: Where the Balgentiens proudly call themselves "cats," in recognition of their wit, in this story, the wronged devil turns that praise into criticism—insulting them by saying in French, "Vous n'êtes que des chats!" ("You're nothing but cats!") Translating *Cat* into French requires an approach to the shifts in languages and cultures that Joyce has brought to the text, as evident in the devil's words here; in particular, the translator faces the inclusion of French in Joyce's telling, the mayor's Irish name, and Joyce's wordplay.

In translating the text into French, the text risks becoming primarily a mirror to France's own town and customs, and eliding Dublin from its implications. To keep the sense that this was not originally so, the French translations explicitly make strange what would naturally be domestic were it a French story. Borel italicizes the French dialogue that Joyce has in the original text, and he adds an endnote: "Les passages en italique sont en français dans le texte originale" ("The passages in italics are in French in the original text"). This note is most likely there not for the sake of its child readers but to remind adult readers that this French text is not an original, at the point when one is most likely to forget this: where Joyce himself wrote French into it. The postscript of Joyce's letter is also included in *Chat*, which enhances the sense of the devil's, and overall of the text's, foreignness and ends the whole story with the word "Dublin":

"il sait aussi parler à la perfection un très mauvais français, quoique ceux qui l'ont entendu assurent qu'il a un fort accent de Dublin" ("he also knows how to perfectly speak very bad French, although those who have heard him say that he has a strong Dublin accent"). In this final sentence, again, it is precisely the French language that is made foreign-sounding, with its strong Dublin accent, despite being a French-language edition of Joyce's story.

The treatment of names in the translations furthers the sense that this French story is not at home in France. Jan Van Coillie tells us that it is particularly uncommon to foreignize names in books for children under ten (135), yet no translator of *Cat* has altered the name "Alfred Byrne." Thus, the translators have to greater or lesser extent, maintained *Cat*'s implicit critique of Dublin. This may have to do with the position that a text like Joyce's holds in the literary polysystem (Even-Zohar, *passim*). Typically, a picturebook would be placed with rather low authority in the literary polysystem, and consequently be open to significant changes in any adaptations or translations; however, since the author of this text is considered to come from High Modernism, translators may be more conscientious to maintain as much of Joyce's language and content as possible in translation. The mirror that Beaugency provides to Dublin in *Cat* is thus replicated in translations, although it inevitably becomes more complex when the story is translated into the language of its original location.

The high literary authority placed upon Joyce's work is also visible in the translation of bilingual punning in the story. The devil tells the people of Beaugency that they are not "belles gens" ("nice people"), which further presses against relations of place and language since this insult echoes the implied positivity of *beau gens* (*sic.*) in their town's name, "Beaugency," claiming that it fails to describe them. As Garnier notes, the *belles* of "belles gens" recur in the name of the language that the devil speaks, "*Bell*sybabble" (101, my italics). This neologism, also found in *Wake*'s "belzey babble" (64.11), evokes a tension of place and language not only in the contrast of *belle* and *beau* but also through its allusion to the biblical Babel. Borel's French translation of this word cannot hold together the homophonic conjunction of *Babel* as a legendary place and *babble* as absurd speech and opts for "diababélien." In choosing *Babel* the place, a high register allusion has been selected over the possible implications that involving *babiller* ("to babble") would have entailed. As with the foreignization of the mayor's name, this is possibly in deference to the erudition one associates with Joyce's language. At the same time, "diababélien" effectively maintains the pleasurable silliness of "Bellsybabble," which Coillie, writing on names in the translation of children's literature, contends is more valuable than seeking to capture all the implications of wordplay (129).

Borel's translation decisions show the complexities of "back" translation since Joyce's use of French becomes a barrier rather than a bridge between texts, his use of an Irish name sits conspicuously in a French folk tale, and his use of neologism forces Borel to decide between a high register allusion or simpler wordplay. The overall effect of these decisions is to foreignize the French language itself in the story and to maintain Joyce's authorial critique of Dublin.

Corre's Cubist and Pastiche Picturebook

For the first French edition in 1966, Jean-Jacques Corre's illustrations are black and white ink prints on high-quality off-white paper in a book that is unusually square. Four beautiful pages near the center of the book are white text and illustrations on black paper—as night falls and the people of Beaugency sleep—for the lines "La nuit vint, tous les gens de Beaugency allèrent se coucher et s'endormirent" ("Night came, all the people of Beaugency went to bed and slept"). More so than Rose's devil, Corre's is not simply Joycean but explicitly a portrait of Joyce's head, made disproportionately large to the rest of his body. As mentioned above, he first positions the face of Joyce mid-sentence, after the words "Le Diable" ("The Devil"). In contrast, the mayor is almost disturbingly faceless, perhaps to figure him as a characterless puppet of authority or bureaucracy. On most pages, the writing extends across the gutter, and letters and short words are often lost in the middle of the two pages. While almost half of the pages have a single line of text or only individual words, a quarter of the pages are overcrowded with text and would most likely be difficult for an early reader. The words often follow the shapes of the illustrations, for instance, with a single line of text serving as a road for people's feet for the lines "si tu voulais la traverser d'une rive à l'autre, il te faudrait bien faire au moins mille pas" ("if you wanted to cross from one riverbank to the other, you would need to take at least a thousand steps"). The text is large and comes at a slant along the bridge for the lines "O Loire, le beau pont!" ("Oh Loire, the beautiful bridge!"), filling a double-page spread. As the cat runs from the water being poured on it, the words jump along its back, following the cat into the devil's arms, and the sentence ends appropriately with its last two words in the bottom right corner beneath the devil: we see the cat crossing the bridge "à toutes pattes, les oreilles rabattues, il vint se jeter dans les bras du diable" ("at full speed, ears turned down, he threw himself into the arms of the devil").

The letter's postscript is given its own double-page spread and printed sideways, so the reader has to turn the book to read it. Above the postscript is an equally sideways image of the devil creating a speech bubble full of pictures, symbols, hieroglyphs, and scattered letters. Its variety of images evokes the all-encompassing nature of the devil's speech, which seems to contain the natural world (owls, crocodiles, fish, cats) and even the supernatural world (witches, a mermaid, a four-leaf clover). Moreover, it connects with the modernist understanding of ideograms and hieroglyphs, as put forward by Ernest Fenollosa and propagated by Ezra Pound, as a kind of language that combines the verbal and the visual (*passim*). This is a particularly apt connection to make because it suggests that the modernist tying of visual to verbal can relate directly to the present picturebook, self-referentially pointing to a possible bridge between modernism and children's picturebooks. Corre's style is itself modernist, conjoining ornate medieval-style manuscript illustrations with cartoon-style caricature, particularly in the giant head and cape of the devil. Medieval architecture takes on a cubist perspective for the houses along the riverbank, directly linking this text with modernist visual art and with an art movement that began in Paris. During the night, the people of Beaugency dream of bridges in various styles, but one baby stands out by anachronistically dreaming of the Eiffel Tower. As such, Corre's illustrations connect with the anachronism, modernity, and child audience of Joyce's text in a specifically French way.

Blachon Drawing Across Cultures

Twenty years after Borel's translation, Joyce's own grandson Stephen published a "new" French translation with his wife. The translation by Stephen and Solange Joyce purports to be entirely different from that of Borel—"Traduction de Jacques Borel entièrement revue par Solange et Stephen J. Joyce pour l'édition de 1985" ("Jacques Borel's translation entirely edited by Solange and Stephen J. Joyce for the 1985 edition")—but it in fact differs in very few places and follows Borel's translation decisions at several key points: the neologism "diababélien" for "Belsybabble," the onomatopoeic expression "le temps de dire ouf, plouf!" ("the time to say ouf, plouf!") for "quick as a thought," and "Le diable piqua une vraie colère de diable" ("The devil flew into a real devilish rage") for "The devil was as angry as the devil himself." As such, Stephen and Solange's apparent dissatisfaction with Borel's translation cannot be seen entirely as a lexical issue. Therefore, it is likely that their decision to retranslate the text is due to the difficult-to-read, but visually beautiful, pagination, and illustration by Corre for Borel's edition. Thus, along with this new translation come the new illustrations of Roger Blachon.

Blachon's illustrations do not simply domesticate or foreignize the mayor and devil but bridge distance and difference in an Irish-French Catholic association. Blachon's work breaks out of the anachronisms of dress that most illustrators employ—making the mayor look like a pope, giving the villagers medieval dress, and designing a medieval town for the backgrounds. In doing so, he envisions similarities between the two cultures—medieval French at the time of the legend and contemporary Irish. Further, he brings out the Joycean theme of Catholicism, which Joyce associates with cats, bridges, and devils across his work. *Wake*, in the simplest example, makes *Catholic* into "Catalick" (158.4), to emphasize the cat link. The mention of the pope as "pontofacts massimust" in *Wake* (532.9) is a twisting of *pontifex maximus*, the Roman Catholic title for the Pope. This translates as "greatest bridge-maker," thereby identifying Catholicism with bridges. Another possible reference to Byrne in *Wake*, alluding to his trip to New York, brings in a particularly cat-like Catholic blessing: "Mon signeur of Deublan shall impart to all, *Benedictus benedicat*!" (569.16–21).[4] Garnier notes that the cat in the "Calypso" section of *Ulysses* "tipped three times and licked lightly" (Joyce, *Ulysses* 54), which she calls a "Trinitarian, Vichian ritual" (Garnier 105). The "Mime" chapter of *Wake* associates bridges with the devil: "oaths and screams and bawley groans with a belchybubhub and a hellabelow bedemmed and bediabbled the arimaining lucisphere, Helldsdend, whelldselse! Lonedom's breach lay foulend up uncouth not be broched by punns and reedles" (239.32–36). While London Bridge is said here to not "be broached by puns and riddles," it is precisely through such wordplay as "hellabelow," "bediabbled," and "lucisphere" that the bridge comes to be understood in all its devilish undertones. Blachon's bridging of Irish and French differences through Catholicism is thus particularly apt for *The Cat and the Devil*, fitting into the matrix of concerns in Joyce's works.

Conclusion

As we have seen, both cats and letters can serve as cross-cultural textual bridges. Joyce's story is itself a translation, domesticating the mayor and devil by giving them Irish and contemporary characteristics in order to mutely set up Beaugency as a critical reflection of Dublin. In translating *The Cat and the Devil* into French, *Le Chat et le Diable* consequently problematizes the relation of place to language—in keeping foreign where one would ordinarily expect domestication and in forcefully estranging its own language. In crossing the bridge once more, we see, one can only return much changed. However, in drawing the story world, Corre and Blachon both envision lasting bridges from the time and space of the author and those of his audience. In connecting modernism and children's literature, Corre enables a child reader (or experiencer) of the book to develop a familiarity with modernist or avant-garde aesthetics. Through Catholicism, Blachon conjoins the two cultures of this story, and his illustrations became a success in not only the target culture but even the source culture for Joyce's story. Where the French translated texts strain domestications into foreignizations, Corre and Blachon's art speak against this move and open the way for a middle ground between these two positions.

Notes

1. Caroline Marie has written about the reception history of Joyce and Virginia Woolf and the market forces that led to the translations of their picturebooks in the 1960s ("Marketing Modernism for Children" *passim*). Here and elsewhere, Marie has written nuanced analyses on the French, Italian, and Australian re-illustrations of Woolf's children's story, which run in interesting parallel with the re-illustrations of Joyce's work.
2. It is common practice when quoting *Finnegans Wake* to do so by page number and line, in the form [page].[line].
3. In some editions, such as the 1990 Breakwater edition published in Canada, "Bellsybabble" is (perhaps mistakenly) changed from Joyce's original wording to be written as "Bellysbabble," which also links it to the Devil's protruding belly in Blachon's illustrations.
4. For more on these links and others between *Cat* and the *Wake*, see Lewis.

Works Cited

Budgen, Frank. *James Joyce and the Making of Ulysses*. Indiana University Press, 1967.

Coillie, Jan Van. "Character Names in Translation: A Functional Approach." *Children's Literature in Translation: Challenges and Strategies*, edited by Coillie and Walter Verschueren. St Jerome, 2006, pp. 123–140.

Dale-Green, Patricia. *Cult of the Cat*. Heinemann, 1963.

Eliot, T. S. *Chats!* [Cats!]. 1939. Translated by Jacques Charpentreau, Arc-en-poche, 1983.

Even-Zohar, Itamar. "Polysystem Theory." *Poetics Today*, vol. 1, no. 1–2, 1979, pp. 287–310.

Fenollosa, Ernest. *The Chinese Written Character as a Medium for Poetry*. 1919. Edited by Ezra Pound. City Lights, 1936.

Garnier, Marie-Dominique. "The Lapse and the Lap: Joyce with Deleuze." *James Joyce and the Difference of Language*, edited by Laurent Milesi. Cambridge University Press, 2003, pp. 97–111.

Hodgkins, Hope Howell. "High Modernism for the Lowest: Children's Books by Woolf, Joyce, and Greene." *Children's Literature Association Quarterly*, vol. 32, no. 4, 2007, pp. 354–367.

Joyce, James. *The Cat and the Devil*. 1936. Illustrated by Gerald Rose. Faber, 1965.

———. *Le Chat et le Diable*. 1966. Translated by Jacques Borel, illustrated by Jean-Jacques Corre. Gallimard, 2009.

———. *Le Chat et le Diable*. 1985. Translated by Solange and Stephen Joyce, illustrated by Roger Blachon. Gallimard Jeunesse, 1990.

———. *Finnegans Wake*. 1939. Penguin, 1992.

———. *Letters of James Joyce: III*. Edited by Richard Ellman. Faber, 1966.

———. *Ulysses*. 1921. Edited by Jeri Johnson. Oxford University Press, 1998.

Lewis, Janet. "*The Cat and the Devil* and *Finnegans Wake*." *James Joyce Quarterly*, vol. 29, no. 4, 1992, pp. 805–814.

Marie, Caroline. "Marketing Modernism for Children; or How Joyce's *The Cat and the Devil* and Woolf's *Nurse Lugton*… Were Made into Picture Story Books in English, Italian, and French." *Leaves*, vol. 5, 2018, pp. 1–22.

Oittinen, Riitta. *Translating for Children*. Garland, 2000.

O'Sullivan, Emer. *Comparative Children's Literature*. Translated by Anthea Bell. Routledge, 2005.

Sigler, Amanda. "Crossing Folkloric Bridges: The Cat, The Devil, and Joyce." *James Joyce Quarterly*, vol. 45, no. 3–4, 2008, pp. 537–555.

Venuti, Lawrence. *The Translator's Invisibility*. Routledge, 1995.

The Translation and Visualization of Tolkien's *The Hobbit* into Swedish, the Aesthetics of Fantasy, and Tove Jansson's Illustrations

Björn Sundmark

Transmedia storytelling is all about piecing together, adding to, embellishing, editing, transforming, and translating. When I think of *The Hobbit*, I have both Tolkien's works (including his own illustrations) and Peter Jackson's films in mind at the same time. I remember a radio adaptation I listened to as a child, as well as an LP with Bo Hansson's electronic music; I think of the annual Tolkien calendars; I think of scholarly and critical work; I think of Pauline Baynes's and Tove Jansson's illustrations; I also think of different Tolkien websites and forums; finally, I think of *The Hobbit* in both Swedish and English. All of these "pieces of Tolkien" are of course rather random and not at all part of a concerted narrative (or marketing) strategy. In practice, however, they become prime examples of transmedia storytelling.

A consequence of adopting a transmedia approach when reading a classic like *The Hobbit* is that other versions, translations, visualizations, and adaptations are allowed to matter and take place. They are acknowledged as part of the whole story, they too become "leafs on the tree of story," to use Tolkien's metaphor from *Leaf by Niggle*. In other words,

B. Sundmark (✉)
Malmö University, Malmö, Sweden

A. Kérchy and B. Sundmark (eds.), *Translating and Transmediating Children's Literature*, Critical Approaches to Children's Literature,
https://doi.org/10.1007/978-3-030-52527-9_7

the transmedia history of a text such as *The Hobbit* becomes an account of how this story has impacted readers and inspired Hobbit-contributors over the years. It also means that if you want to understand the original text and context better, these other texts turn out to be significant. In this chapter, I will show how changing perceptions on how Tolkienian fantasy should be visualized can be traced through the different approaches to illustrating *The Hobbit* and *The Lord of the Rings*. My main example will be Tove Jansson's 1962 illustrations for the second Swedish edition of *The Hobbit*.

Translation and Illustration

Translation, like transmediation, involves reinterpretation, and it provides new versions of a source text (or several). A translated text does not "require" re-mediatization. However, translation also goes hand in hand with sometimes drastic changes in the format, design, quality (paper, binding, etc.), and use of extratextual resources (prefaces, annotations, blurbs, etc.). The inclusion of new artwork to accompany a translated text represents a significant change from the source text—and a re-mediatization in its own right. Illustrated books present specific challenges for the translator. If there are illustrations, the translator must make sure that references to the illustrations are not lost or contradicted in translation (Lathey 55). Moreover, illustrations are often altered, deleted, or placed differently in translations, which affects the overall impression and effect of the text (Stolt). It would seem ideal that illustrations can be carried over unchanged from one language to another, whereas text always has to be translated. However, in cases where the translation is an attempt at modernization and adaptation to the target language, the original illustrations may seem oddly out of place. For instance, the effect of a verbally modernized, Swedish-speaking *Alice in Wonderland* in the company of Tenniel's illustration can be slightly jarring. The illustrations can easily be perceived as bizarre, exotic, and disconcerting, rather than beautiful, precise, and congenial. (This is partly true of English editions as well. In most cases Carroll's language seems less dated and more open to interpretation than Tenniel's iconic illustrations—a situation which goes some way toward explaining why many English Alices also make use of more modern illustrations.)

A problem of a different order is when the illustrations in the original work are replaced by new illustrations (Stolt), or when the original work is unillustrated. In the case of *The Hobbit*, the first edition carried ten black and white illustrations and two end-paper maps by J. R. R. Tolkien himself; the author also supplied the cover and dust-jacket illustration. However, none of Tolkien's original illustrations, not even the maps, were reproduced in the Swedish 1962 edition. Instead, the Tove Jansson *Hobbit* has eleven full-page illustrations, black and white illustrations, and thirty six small illustrations/vignettes, as well as Tove Jansson's cover. An analysis of the translation and transmediation of Tolkien's 1937 *The Hobbit* that resulted in the 1962 Swedish *Bilbo—en hobbits äventyr* will therefore have to account for the discrepancy between the two editions with regard to the illustrations while discussing the underlying artistic and literary choices made. Similarly, the subsequent rejection of Tove Jansson's illustrations in favor of Tolkien's represents yet another shift in the sensibilities of publishers and readers with regard to the changing aesthetics of fantasy.

Tolkien on Fantasy and Art/Illustration

Tolkien did not only write and illustrate *The Hobbit* and many of his other texts about Middle-Earth, he also theorized about the role of art and illustration. His critical views—not only of his work but of fantasy in general—are important since they continue to influence artists, publishers, and readers. In his essay "On Fairy-Stories," he asserts the primacy of words over illustrations: "In human art Fantasy is a thing best left to words, to true literature" (45). In the accompanying endnote E, he goes on to say that,

> However good in themselves, illustrations do little good to fairy-stories. The radical distinction between all art (including drama) that offers a *visible* presentation and true literature is that it imposes one visible form. Literature works from mind to mind and is thus more progenitive. It is at once more universal and more poignantly particular. If it speaks of *bread* or *wine* or *stone* or *tree*, it appeals to the whole of these things, to their ideas; yet each hearer will give to them a peculiar personal embodiment in his imagination. (67)

Nevertheless, as we know, Tolkien *did* illustrate his own work. How should one understand this? John R. Holmes traces Tolkien's views on the illustration of fantasy to the pre-Raphaelite artist and writer William

Morris—admired by Tolkien—but also recognizes in Morris' writings an opening toward the visual expression of fantasy that could explain Tolkien's own illustrations. Morris asserts that for "decoration" to be of any value, it must "remind you of something beyond itself, something of which it is but a visible symbol" (quoted in Holmes 28). This suggests an art that is stylized and symbolic rather than realistic and particularized. There is also the underlying intimation that the art should be expressive in the sense of evoking ideas and emotions rather than depicting specific moments and places. Furthermore, even when pictorial art does succeed in rendering something beyond itself, as in surrealism, it is all too often given to grotesque and morbid effects according to Tolkien (67). In a letter to Rayner Unwin (23 May 1961) discussing the possibility of finding an illustrator for *The Lord of the Rings*, Tolkien writes pessimistically, "it is impossible to hope, nowadays, that one might come across an artist of talent who could, or would even try to depict the noble and heroic" (quoted in Blok 25). Of course, one reason why it was so hard to find illustrators who were up to the task was that the noble and heroic in this period was associated either with juvenile superheroes like Batman and Superman and heroic Hollywood stereotypes or with the tainted iconography of the Nazi, Fascist, and Communist regimes of the twentieth century.

Now, if we turn from Tolkien's criticism to his practice, we see that his illustrations for *The Hobbit* indeed are largely symbolic and stylized, even when they depict a specific moment in the story. It is also clear that he prefers atmospheric landscapes—*Stillebens* on a large scale, so to say—to actions and individual characterization. As Hammond and Scull have pointed out,

> [even] the few illustrations of particular scenes in *The Hobbit* are more notable as settings than for what is going on within them. Tolkien provided backgrounds on which readers can paint their own mental pictures, directed by a text but not constrained by too specific an image. (98)

One can see this as a strategy to make illustrations that point beyond themselves, toward fantasy, in line with the view presented in "On Fairy-Stories." At the same time, Tolkien's illustrations, just like his writing, tend toward verisimilitude and "realism of presentation," to use C. S. Lewis's term. By implication, the descriptions tend to be precise and correct. Moreover, the illustrations must not contradict or expand significantly on the verbal text; proportions and scale must not challenge reason. After all, Middle-Earth is not a completely different place but rather an alternative (or earlier) version of the one we inhabit. Thus, the impulse toward symbolism and the impulse toward realism of presentation tend to balance out in Tolkien's work. The dust-jacket illustration depicting the Lonely Mountain on the first (and many subsequent) edition and the two end-paper maps are satisfying on both counts; they open up for interpretation and give the reader the impulse to visualize the fantasy further while they anchor the fantasy in a consistent and made-believable place (see Sundmark, "Mapping").

As for other visual interpreters of his work, Tolkien showed sympathy for Pauline Baynes' illustrations; he writes that they are "akin to his own" and have "a touch of fantasy'" (Carpenter 350, 312). Still, his praise was not unreserved; her illustrations apparently fell short of his standards for the noble and heroic. Nevertheless, Baynes remains the most important Tolkien illustrator in English. She made the Puffin cover for *The Hobbit* (1961) and the iconic one-volume paperback cover of *The Lord of the Rings* (1968),[1] as well as the poster maps for both of these books. Baynes also made the covers for *Farmer Giles of Ham* (1949), *The Adventures of Tom Bombadil* (1962), *Smith of Wootton Major* (1967), and *Bilbo's Last Song* (1974). Other than Baynes, Tolkien also appreciated the work of the Dutch artist Cor Blok and a few others, such as Mary Fairburn. However, none of them were featured in English editions of his work during his lifetime, nor were anyone else (for Tolkien's views on different artists' visions of his work, see Hammond and Scull; Davis; Liptak; Blok; Tankard). Until recently, Tolkien's own in-text illustrations for *The Hobbit* and cover designs for *The Hobbit* and *The Lord of the Rings* have held an almost hegemonic position in English-language publications. Besides the work of Pauline Baynes, the only significant exceptions to the rule of Tolkien are the covers of the two 1965 paperback ventures of *The Lord of the Rings* in the United States: Jack Gaughan's horror-inspired cover for ACE and Barbara Remington's "emus and Christmas-trees" cover for Ballantine. After Tolkien's death in 1973, we gradually start seeing more media and genre diversity, as well as alternative visualizations of Middle-Earth—*The Tolkien Calendar* (1973–), Ralph Bakshi's film venture (1978), fan art, illustrated books, and so on. Nevertheless, there is much to suggest that the Tolkien fandom also exerted a homogenizing and conserving influence on what was acceptable and publishable for a long period of time.

Illustrating *The Hobbit* and *The Lord of the Rings* 1947 to 1973

It is to the early translations of *The Hobbit*[2]—and a lesser extent *The Lord of the Rings*—we must turn if we want to understand how Tolkienian fantasy was understood and visualized up to 1973 (during Tolkien's lifetime, that is) and how varied the response was to his work (see Anderson). Between 1947 and 1973, *The Hobbit* was translated into fourteen languages in sixteen different editions: Swedish (1947 and 1962), German (1957, 1971), Dutch (1960), Polish (1960), Portuguese (1962), Spanish (1964), Japanese (1965), Danish (1969), French (1969), Norwegian (1972), Finnish (1973), Italian (1973), and Slovak (1973). Nine of these editions were fully illustrated by eight illustrators, since Tove Jansson's work was used for both the Swedish and Finnish publications. Three of the remaining publications only featured original cover art (the Dutch, Spanish, and Norwegian editions), while the rest used work by Tolkien. In the same period, there appeared ten nearly complete translations of *The Lord of the Rings*: Dutch (1957), Swedish (1959–1961), Polish (1961–1963), Italian (1967–1970), Danish (1968–1972), German (1969–1970), Japanese (1972–1975), Finnish (1973–1975), and Norwegian (1973–1975). All of these feature original cover art. The Danish publication also contains original in-text illustrations. The chronology shows that in many cases the first publications of *The Hobbit* and that of *The Lord of the Rings* followed each other closely in time. It seems likely that the decision to publish the longer, later, and by then world-famous work may have prompted the decision to also translate and publish *The Hobbit*. One can see that this might have complicated the illustrators' task. If read as a prequel to the more adult *Lord of the Rings*, it should presumably harmonize with the atmosphere, tone, and generic conventions of that work. An even greater challenge must have been the perceived genre indeterminacy of both works. For one should keep in mind that "fantasy" was not a publishing category or an identifiable genre (adult or child) at the time. Tolkien's "fairy-stories" helped establish fantasy as a genre, certainly, but it would be anachronistic to expect readers (including illustrators) in the 1950s, 1960s, and 1970s to see it as such. It could conceivably be treated as a fairy tale, an adventure story, an alternative history, or myth.

Most of these early illustrators choose to emphasize that *The Hobbit* is above all a humorous children's book—maybe as a way of balancing the more frightening parts but perhaps, too, because other genre markers were difficult to pick up on. In general, the illustrations are stylized, decorative, naïve, and expressive. They are not realistic, and they do not idealize or romanticize the story. For example, Jan Mlodozeniec (Poland) and Antonió Quadro (Portugal) both opt for comic book-styled innocence and comedy. Horus Engels and Klaus Ensikat (Germany) aim for child-oriented humor, but through caricature and exaggeration. Ensikat is also interesting because of his use of historical elements in dress and architecture, something seen in many later Tolkien illustrators. The visual humor of Torbjörn Zetterholm (Sweden) and Nada Rappensbergerova (Slovakia) is more grotesque and dark, visually as well as in tone, but still with a child audience in mind. Finally, Ryuichi Terashima (Japan) and Tove Jansson (Sweden, Finland) emphasize the frightening and dramatic aspects of the text; they are arguably the least childish in their approach of the pre-1973 illustrators. Like Tolkien, they also display a keen interest in the setting—the high mountains, the deep forests—but add a dynamic and expressive dimension to their work absent in Tolkien's illustrations. None of these eight early Hobbit-illustrators attempts anything like Barbara Remington's psychedelic-surrealistic 1965 illustration. Nor can we see anything in the artwork that presages the "realistic" high fantasy mode of illustration that dominated throughout the 1980s and 1990s (for reference, see the *Tolkien Calendars* for the period or the two anonymously edited collections *Tolkien's World* and *Realms of Tolkien*).

The covers for the translated *Lord of the Rings* pre-1973 are also noteworthy. Some of them—like the Dutch (1965), Italian (1967), Japanese (1972), and Finnish (1973)—represent a naïve or child-oriented aesthetics that harmonize well with the corresponding artwork for *The Hobbit*. These covers do not signal an adult audience. At the other end of the spectrum, we find the austere French cover (1972), which looks more like the cover of a scientific journal than that of a work of fiction—only text (author, title, and publisher) against a blue background. The strict cover design of the Danish edition also suggests an adult readership. The German cover (1969) likewise signals sophistication and an adult audience, but in this case, by choosing a non-realistic, experimental mode of illustration (the same can be said of the German editions in 1979 and

1984). By contrast, the Dutch (1956) and the Polish (1961) covers strike a balance: instead of scenes and characters, they represent symbolic things and designs; they are rather like Tolkien's own dust-jacket designs for *The Lord of the Rings* (the ring, the eye, and elvish lettering). The Swedish covers by Rolf Lagerson, both the ones for the first edition 1959–1961 and for the 1967 paperback edition, also work as crossover-illustrations; they represent identifiable scenes and actions from the books, yet they do so in a non-realistic and highly stylized form. The first three (1959–1961) draw inspiration from Greek designs and paintings, while the 1967-set is less stylistically determined.

All in all, while the pre-1973 illustrations and covers are often stylized and suggestive, they do not strive for realism of representation, nor do they appear to make any overt use of symbolism. Instead, the illustrators focus on the child reader (as we can see, this is also true of some of the covers for *The Lord of the Rings*) and bring out the humorous aspects of *The Hobbit*. They also struggle—often very creatively—to visualize what kind of book it is, that is, the genre (fairy tale, horror, adventure story, or myth/legend). Inadvertently, these early illustrations represent the genesis and development of fantasy illustration. I will now examine more closely the illustrations and reception of one of these pioneers of fantasy illustration, Tove Jansson.

Tove Jansson's *The Hobbit*

The Swedish 1962-edition of *The Hobbit* was undoubtedly prompted by the 1959–1961 publication of *The Lord of the Rings*. By then, the first Swedish edition of *The Hobbit* (1947)—the first in any language—was long out of print, and Tore Zetterholm's translation had drawn criticism (Tolkien himself disliked the translation of "hobbit" into "hompe"). The illustrations in that first edition were also seen as wanting and childish, especially in the context of the phenomenal success of the Swedish translation of Tolkien's adult fantasy. For the purpose of coming up with "the children's book of the century," Astrid Lindgren, who worked as an editor at the publishing house Rabén & Sjögren, teamed up with Britt G. Hallqvist and Tove Jansson. Hallqvist was one of the most respected translators of the time, having made notable translations of both children's and adult literature (Rudyard Kipling, Edward Lear, T. S. Eliot, Laura Ingalls Wilder, and many more). What Hallqvist and Jansson (and Lindgren) had in common was that they appealed to a cross-generational audience in their work. This was surely intentional. Moreover, unlike the 1947-edition, the book was produced in octavo format on high-quality paper. Thus, the second edition of *The Hobbit* was a high-end project that could match the emerging status of *The Lord of the Rings* as a classic.

As I have argued above, the early illustrations of *The Hobbit* are part of a translation and transmediation process. As we shall see, Jansson's role is particularly interesting in such a context. The all-encompassing "Moominverse" that she created over several decades is in itself an excellent example of transmedia storytelling: The comprehensive Moomin-narrative comprises nine illustrated books (one of which is a short-story collection), a number of picturebooks, a comic strip (1954–1959),[3] a Moomin house, Moomin dolls, films, a theme park, and so on. Jansson herself worked in, with, and across different media. As an illustrator of someone else's work, she would naturally bring that transmedia sensibility to bear on it.

It was around the time when Jansson's Moomin involvement peaked in the late 1950s and early 1960s that she accepted to illustrate three classic works of children's literature: Lewis Carroll's *The Hunting of the Snark* (1959) and *Alice in Wonderland* (1966), as well as J. R. R. Tolkien's *The Hobbit* (1962). The commissioned work gave Jansson the opportunity to venture beyond Moominvalley, to receive new impulses, and to develop her technique in new ways. Boel Westin writes that Jansson wished that "the Tolkien illustrations should speak in their own right, as an expression of the artist Tove Jansson, not Tove, the Moomin-illustrator" (348). The vignettes in particular became a way of distancing herself from her regular "Moomin-style" (349). However, it turned out that the "studious spontaneity" Jansson developed for the vignettes in *The Hobbit* was a technique she would adopt in her later Moomin books (Holownia 4–5). Thus, while her artwork for *The Hobbit* represents something new—technically, but also because it represents another writer's imaginary universe—it is still undeniably "Jansson," and anyone familiar with her earlier illustrations can see that. One can also argue that her later Moomin books have been influenced by her work on Tolkien and Carroll. In fact, one understands how their worlds and words must have resonated with hers. Like Jansson, Carroll and Tolkien are (in their different ways) fantasy writers as well as thinkers, and their books appeal to children and adults alike. We see this in Jansson's work as well: the Moomin books are for all ages, and the later ones in particular blur the line between child and adult reader. Jansson was certainly heading in this direction already

(Westin, *Familjen i dalen*), but it is likely that the Carroll–Tolkien experience helped her on the way, both as an illustrator and as an author, and that it facilitated her exploration of darker and more complex themes in her work.

For Jansson, the main attraction of *The Hobbit* was that, just like *The Hunting of the Snark*, it had potential for "horror." Jansson deemed Tolkien "more gruesome than Poe," which was high praise coming from her (Westin 346). Elsewhere, too, Jansson talks with relish about "forests of living terror, coal-black rivers, moonlit moors with fiery wolfs—a whole world of catastrophe that I know I can respond to in pictures" (quoted in Westin 348). She does not talk about it as children's literature, nor as being humorous, nor as a fairy tale—except rather disparagingly, when commenting on Tolkien's stock characters and creatures: "The characters are commonplace: dwarfs, goblins, elves, black demigods"[4] (ibid). In consequence, the main characters are generally decentered and downsized, placed in the margins, or used as vignettes, thereby highlighting the grandeur of the landscape and the diminutive role of those that people it. Olga Holownia writes of Bilbo and company,

> They become almost spectators rather than participants in their quest. In all images, the framing of which usually conveys confinement, we find the dwarves and Bilbo hidden among the branches way above the dancing goblins and wargs, wandering among the overwhelming trees of Mirkwood and equally small on the bridge to Rivendell. They are literally on the edge of the picture in the scene showing their approach to the Lonely Mountain. (10)

An exception to the rule, and the only time Bilbo is allowed to dominate visually, is on the cover. However, this is only because the original design was rejected; Jansson's first cover showed the company ascending a mountain pass and the dragon dominating the sky above them.[5] However, the publisher (and Lindgren) regarded this cover as too adult and frightening for children, and she persuaded Jansson to make a new one (Westin 349). The result unfortunately jars with the overall conception of the artwork, and it has been roundly criticized for showing Bilbo playing dress-up with a halberd and shield and a dragon engaged in choreographic flight in the distance (Sundmark 10). Even the background color—pink!—seems intentionally childish.

If the main characters are generally represented as small and insignificant, the goblins, wargs, trolls, and the like by contrast are often disproportionally large, occupying center stage: the trolls that tower over the forest (48), or the giant, looming Gollum (85). In this respect, the monsters are treated as part of the powerful and antagonistic setting, and vice versa. The landscape itself is monstrous. Commenting on the illustration showing the dwarves' approach to the Lonely Mountain (209), Holownia observes,

> [W]e do not see the dragon but we are made aware of his "disguised" presence. The atmosphere of impending danger is heightened by the horizontal lines of the flames and a cloud of steam coming from the cave over the waterfall. This alignment disrupts the otherwise vertically designed scene (steep rocks, a waterfall, tall mountains). And while the four tiny figures are marginal observers, one can imagine the whole landscape as being one enormous monster—the cave forming a giant gaping mouth, with a waterfall tongue lolling over sharp, rocky teeth. (11)

Another dramatic landscape scene is when the company is caught in a storm when they are attempting to cross the Misty Mountains (65). Incidentally, this illustration shares the same basic design as the first cover, discussed above. In the full-page illustration, we see Bilbo in the lower right-hand corner, precariously poised and arms outstretched on a jutting mountaintop. There are thunder and lightning, wild clouds, tumbling rocks, and in the distance, lights from a lakeside town. Again, a small character is "cornered" and contrasted with raging elements and vast spaces. However, this illustration can also be used as an example of Jansson's expressive and non-realistic style. There is no attempt at verisimilitude; instead, it is sound and fury, motion and emotion, *Sturm und Drang*. Moreover, the illustration shows things that are not mentioned in Tolkien's text, like the glimpsed lights from the distant town. For readers who want fidelity to Tolkien above anything else, this may be disturbing; for others, it makes Middle-Earth a larger and more interesting place.

As we have seen, Jansson is mostly interested in Tolkien's settings. Unlike Tolkien's own illustrations, she uses characters and creatures to make the scenes come alive. The characters themselves are usually undeveloped and sketchy and are there mainly to create contrast and movement. Moreover, the sketchiness of Bilbo, especially, means that he is open for interpretation and makes him easy to imagine or identify with. Another aspect of Jansson's characters is their dynamism. In contrast to Tolkien's rather static illustrations, Jansson excels in movement and energy. Finally, even though Jansson herself downplays her ambition for characterization, some of her portraits (like the one of an unrepentant Thorin in the Elvenking's dungeon) are both suggestive and perceptive (174).

If at all, Tolkien wanted illustrations of fantasy to be stylized and symbolic: they should evoke ideas and emotions and point to something beyond themselves. However, part of his secondary world-construction also relies on creating a semblance of a real world, of achieving verisimilitude. Jansson's illustrations accomplish the first, not the second; she does not strive for verisimilitude or verbal–visual correspondence. Instead, her Hobbit-illustrations are symbolic, expressive, and dynamic. They add to and enrich Tolkien's original vision. Moreover, the world conjured up by her images is not the generic fairy tale nor the (historical) adventure story—it is something different altogether. Together, Tolkien's words and Jansson's images show Middle-Earth as a new creation, a new secondary world. Moreover, by stressing the horror aspect and crossover potential, the affective and aesthetic gap between *The Hobbit* and *The Lord of the Rings* is partially closed. Of course, there is always the risk, according to Tolkien, that horror descends into "silliness" or becomes "macabre and grotesque." In the end, this is a matter of one's own aesthetic judgment. As we have seen above, Tolkien was skeptical about finding any contemporary illustrator who could do justice to the "noble and heroic" (in *The Lord of the Rings*). However, I would argue that several of Jansson's illustrations do precisely that. A striking example is the scene when Bard is taking aim at Smaug (252). The town is going up in flames, people are throwing themselves into the water, boats are capsizing, and Smaug dominates the upper part of the illustration, looking down on the smoking ruins, spewing out fire. The immense dragon does not even fit into the picture completely, and his ponderous weight threatens to crush the town below him. Yet, in the middle of the picture, we see an undaunted miniature, a stick man with a bow. As I have said before, the main characters are almost never centered in Jansson's illustrations, but Bard, the dragon slayer, is at the cross-hairs, literally and figuratively. His heroism is made greater because he is shown to be so very small, and because the forces of destruction and despair are so overwhelming. In that image, we see not only Bard's heroism centered but Bilbo's, too, as well as that of all small, good creatures.

There and Back Again

Curiously, the Jansson Hobbit was short-lived. Despite its qualities, the publication soon vanished from circulation and was replaced by other editions, featuring Tolkien's own illustrations.[6] Thus, few people who started reading Tolkien in the 1970s and 1980s would have seen her work, even in Sweden. Outside Sweden and Finland, her Hobbit-illustrations are even less known. To date, there has been no English edition of *The Hobbit* featuring her work. With the exception of a few examples of her work in Douglas Anderson's *The Annotated Hobbit*, it was not until the 2016 *Tolkien Calendar* that a selection of Jansson's Hobbit-illustrations were made available to an English/international audience—more than fifty years after the original publication (They had of course been available on the internet well before that.) The reasons behind its disappearance are not self-evident. Negative reviews and hostile responses from the Tolkien fandom have been cited (Westin 351), but as I have shown in a previous article, Jansson's Hobbit-illustrations were favorably reviewed by a majority of Swedish critics and "Tolkienists" (members of the Tolkien fandom).

There is much to suggest that although the reading public (including the reviewers) were enthusiastic about *The Hobbit*, they were uncertain about what they were reading and seeing—and thus what to expect of the illustrations. Was it a children's fairy tale, a humorous adventure story, a myth, or a tale of horror? Moreover, most Swedish readers at the time would have read *The Lord of the Rings* prior to *The Hobbit*, which prompted some reviewers to read *The Hobbit* as a (too) childish prequel to what one reviewer expressed as Tolkien's "myth of good and evil." Against such an adult and serious "horizon of expectation," *The Hobbit* falls short simply because it is children's literature. The mere presence of illustrations—an established marker of childishness in literature—corroborates such a reading. In other words, the problem is less to do with the illustrations than with the uncertainty of genre belonging and audience status of *The Hobbit* itself.

Moreover, the 1960s and early 1970s were, as I have shown in this chapter, a period of transition when (Tolkienian) high fantasy emerged as a publishing genre in its own right, eventually bringing with it a quasi-realistic mode of representing the fantastic and an emerging orthodoxy when it comes to how places and creatures "should" be represented. Today, we "know" what an elf or an orc should look like; we did not in the 1950s and 1960s.[7] This development jarred with the expressive and non-realistic artistry of some of the early Tolkien illustrators. Jansson was no exception. However, as the realistic mode of high fantasy was brought to its logical conclusion with Peter Jackson's cinematic visualization of Middle-Earth, we started seeing a reaction against it from different quarters, such as the featuring of Cor Blok (2012) and Tove Jansson (2016) in the *Tolkien Calendar*, albeit half a century after their inception. This is heartening. We are back where we started. Everything is possible again.

Jackson's films present "one vision to rule us all," something that should be quite anathema to Tolkien's own view of the possible role of fantasy illustration, including dramatic representations. The only way to counter the influence is to experience different, powerful visual representations of his work. In the preface to the 2012 *Tolkien Calendar*, Ruth Lacon writes,

> A single visual version of *The Lord of The Rings* has become massively dominant, with no real challenge standing against it. I refer, of course, to the motion-picture version of *The Lord of The Rings*. The "single visual image" of the films creates what I would call visual allegory, a one-to-one correspondence, matching J. R. R. Tolkien's description of verbal allegory in the Foreword to the second edition of *The Lord of The Rings*. Once such an image-set exists there is no way to pretend it does not. The genie is out of the bottle; with satellite broadcast, even an alien from outer space is now likely to have seen *The Lord of The Rings* before they read it. Our only way to undo that "single image" constraint now, I would submit, is not to refuse depiction but to unleash it.

It is through the early-illustrated translations of *The Hobbit* that the development of Tolkienian fantasy aesthetics is made visible. By returning to some of the early illustrations of Middle-Earth, such as the ones created by Tove Jansson, we can recover some of the original openness and imaginative potential that Tolkien invested in his creation. The key to doing so lies in the recognition of translation and illustration as driving forces integral to transmedia storytelling.

Notes

1. The cover was based on her 1963 slip-case design for the trilogy.
2. All the covers of the early translations of *The Hobbit*, as well as all in-text illustrations of *The Hobbit*, can be easily accessed online, for instance, at *Babel Hobbits* http://tolkien.com.pl/hobbit/index.php.
3. Tove Jansson's brother, Lars Jansson, continued doing the Moomin comic strip until 1975.
4. Jansson uses the word *svartalfer* (literally, "black elves"), an unusual word employed in the first Swedish translation, but not by Hallqvist. In all probability, Jansson had read Tore Zetterholm's translation rather than Hallqvist's manuscript version. This hypothesis is further supported by some illustrations—the dwarves smoking, Bilbo meeting Gollum, an eagle carrying a dwarf—which are all conceptually close to the corresponding images in the first edition. Furthermore, Jansson was not familiar with Tolkien's illustrations and would, therefore, hardly have read *The Hobbit* in English while preparing her own work.
5. Jansson's original cover was used on the first Finnish edition of the work *Lohikäärmevuori* (1973).
6. There has been one subsequent Swedish edition (1994), as well as the Finnish 1973-edition.
7. Notably, Tolkien's own verbal descriptions are surprisingly vague, incomplete, or contradictory. Thus, early illustrators (before the emergence of a visual orthodoxy) had very little to go on when visualizing a hobbit or an elf or an orc, or even the landscape itself. See Agøy's "Vague or Vivid? Descriptions in *The Lord of the Rings*."

Works Cited

Agøy, Nils Ivar. "Vague or Vivid? Descriptions in *The Lord of the Rings*." *Tolkien Studies*, vol. 10, 2013, pp. 49–67.

Anon. *Tolkien's World: Paintings of Middle-Earth*. HarperCollins, 1992.

———. *Realms of Tolkien: Images of Middle Earth*. Harper, 1996.

Anderson, Douglas. *The Annotated Hobbit*. Unwin, 1988.

Auger, Emily. "The Lord of the Rings' Interlace: Tolkien's Narrative and Lee's Illustrations." *Journal of the Fantastic in the Arts*, vol. 19, no. 1, 2008, pp. 70–93.

Blok, Cor. *A Tolkien Tapestry: Pictures to Accompany the Lord of the Rings*. HarperCollins, 2011.

Carpenter, Humphrey. *The Letters of J. R. R. Tolkien*. Houghton Mifflin, 2011.

Davis, Lauren. "The Middle Earth Illustrators J.R.R. Tolkien Loved—And the Ones He Abhorred." *io9*, 2012. https://io9.gizmodo.com/5968792/the-middle-earth-illustrators-jrr-tolkien-lovedand-the-ones-he-abhorred.

Drout, Michael D. C., editor. *J. R. R. Tolkien Encyclopedia: Scholarship and Critical Assessment*. Routledge, 2006.
Hammond, Wayne G. and Christina Scull. *J. R. R. Tolkien—Artist and Illustrator*. HarperCollins, 1995.
Holmes, John R. "Art and Illustration by Tolkien." *J. R. R. Tolkien Encyclopedia: Scholarship and Critical Assessment*, edited by Michael D. C. Drout. Routledge, 2006, pp. 27–32.
Holownia, Olga. "'Hell, What a Chance to Have a Go at the Classics': Tove Jansson's Take on *Alice's Adventures in Wonderland, The Hunting of the Snark,* and *The Hobbit*." *Barnboken*, vol. 37, no. 2, 2014.
Lacon, Ruth. "To Illustrate or Not to Illustrate, That Is the Question…" *Tolkien Library*. 2011. http://www.tolkienlibrary.com/press/1026-To-Illustrate-or-Not-to-Illustrate.php.
Liptak, Andrew. "The Unauthorized *Lord of the Rings*." *Kirkus Reviews*, 2013. https://www.kirkusreviews.com/features/unauthorized-lord-rings/.
Sliwinski, Marek. *Babel Hobbits: Gallery of Illustrated, Multilingual Editions of The Hobbit by J. R. R. Tolkien*. http://tolkien.com.pl/hobbit/index.php.
Sundmark, Björn. "'En hobbit och ett mumintroll skulle kunna mötas i bästa sämja': Receptionen av Bilbo, en hobbits äventyr (1962)." *Barnboken*, vol. 37, no. 2, 2014.
———. "Mapping Middle Earth: A Tolkienian Legacy." *Maps and Mapping in Children's Literature: Cityscapes, landscapes, and seascapes*, edited by Nina Goga and Bettina Kümmerling Meibauer. Amsterdam: John Benjamins, 2017.
Tankard, Paul. "'Akin to My Own Inspiration': Mary Fairburn and the Art of Middle-earth." *Tolkien Studies*, vol. 14, 2017, pp. 133–154.
Tolkien Calendars. http://www.tolkiencalendars.com/index.html.
Tolkien, J. R. R. *The Hobbit*. 1937. New York: Harper Collins, 1998.
———. *Bilbo—en hobbits äventyr*. Translated by Britt G. Hallqvist. Illustrated by Tove Jansson. Stockholm: Rabén & Sjögren, 1962.
———. *The Lord of the Rings*. 1954–1955. George, Allen & Unwin, 1968.
———. "On Fairy Stories." *Tree and Leaf*. George, Allen & Unwin, 1970.
Westin, Boel. *Familjen i dalen: Tove Janssons muminvärld*. Bonniers, 1988.
———. *Tove Jansson: Life, Art, Words*. 2007. Translated by Silvester Mazzarella. Sort of Books, 2014.

The (Im)Possibilities of Translating Literary Nonsense: Attempts at Taming Iconotextual Monstrosity in Hungarian Domestications of Lewis Carroll's "Jabberwocky"

Anna Kérchy

Nonsense Literature, an Untranslatable Genre?

Literary nonsense is a genre filled with language games and grounded in the strategic destabilization of coherent meanings, conventional interpretive strategies, and logical reasoning. It bridges the gap between child and adult audiences by offering the latter a temporary retreat from disciplined discourse's structures of authority and by allowing the former a ludic revelry in nonverbal, acoustic registers of signification. It is like a "recess bell that officially freed [youngsters] from the classroom to the playground in their reading" (Darnton in Goldwaite 74), while it simultaneously allowed grown-ups to return to a childhood state, granting them a joyous forgetfulness they wished to understand by rational means. Apart from its trademark neologisms' semantic incongruities, it is this manifold ambiguity—resulting from the dual address, the epistemological challenge, and the oscillation between unspeakability

A. Kérchy (✉)
University of Szeged, Szeged, Hungary

A. Kérchy and B. Sundmark (eds.), *Translating and Transmediating Children's Literature*, Critical Approaches to Children's Literature,
https://doi.org/10.1007/978-3-030-52527-9_8

and over-verbalization—that relates nonsense literature to the notion of the untranslatability.

I find two contemporary French semioticians' complementary viewpoints on making sense of nonsense particularly useful in tackling the challenge of translating the presumably untranslatable. On the one hand, Jean Jacques Lecercle suggests that the ultimate challenge for the translator resides in literary nonsense's language and its philosophically charged stylistic bravado. Nonsense is "always already translated, being made up of meaningless pseudo-words" (Lecercle "Translate it, Translate it not"), which are instrumental in eliciting the audience's immersion in a topsy-turvy fictional reality, but they also stimulate a metanarrative self-awareness concerning discourse's insufficiency in representing reality (*Philosophy of Nonsense*). They disclose the inevitability of misunderstanding and the impossibility of meaninglessness while enacting the contradiction between verbal chaos and verbal constraints. On the other hand, according to Julia Kristeva, nonsense's affective, sensorial, musical charge foregrounds the "revolutionary poeticity" of language use, a rhythmicality that makes the genre so pleasurable for child readers/listeners dwelling closer to the primal corporeal experience of the "semiotic realm" preceding the symbolization integral to the disciplinary socialization process.

Lewis Carroll's Victorian nonsense poem "Jabberwocky"—embedded in *Through the Looking Glass and what Alice Found There*, the 1871 sequel to his fairy-tale fantasy about Alice's adventures in Wonderland (1865)—has been referred to as "the holy grail of translation" (Heyman in Sundmark, "Some Uffish Thoughts" 43). It represents an ultimate challenge since the translators must find a balance between (Lecercle's) metalinguistic philosophical and (Kristeva's) transverbal, acoustic vocal layers of signification—as well as between the logical and linguistic impossibilities, between accommodation to the target audience and faithfulness to the original—while seeking to transpose into another culture the source text's nonsensical *effect*. Moreover, our initial glimpse at the Jabberwock as a fictitious beast of Wonderland's mythology induces what iconologist W. J. T. Mitchell would call an "imagetextual encounter." Alice is first perplexed by the monster's representational mode in mirror writing, a defamiliarizing transcription of words more reminiscent of image than text, while readers' "pleasurable confusion" is enhanced by John Tenniel's by now iconic woodcut engraving of the "un-image-inable" grotesque creature in the first Macmillan print and many succeeding editions. Hence, the translator's task is further complicated by the transmedial extension of the interpretive challenge: Besides the (meta)semantic/syntactic and phonetic acoustic layers of meaning formation, the work's visual components must be taken into consideration throughout adapting the poem to a different language.

My aim in this chapter is to explore how the interaction between verbal narrative and visual illustration, between oral performance and written transcription, and between source text and translation can be regarded as vital constituents of the complex signification process called transmedia storytelling. Henry Jenkins defined transmedia storytelling as a strategic expansion of a fictional universe over a variety of different media platforms and communication channels that each make unique contributions to the unfolding of the storyworld while augmenting new dimensions of the immersive and coordinated entertainment experience. Pictorial, acoustic, and cross-linguistic interpretations may enhance the source-textual meanings in a variety of ways, mutually formative of each other and their varieties. I wish to analyze here the (im)possibilities of translating Jabberwocky on three levels. First, I comment on the Carrollian text's metafantastic/metanarrative reflection on the reader as a translator/interpreter who will inevitably both decode and reproduce/regenerate nonsense—both on the level of implied readership (as in fictional Alice's quest for meaning) and the actual lived interpretive experience of the tale's putative reader or listener. Second, I focus on illustration as a translation of the written narrative and the (author/translator's and illustrator's) self-conscious use of iconotextual dynamics. I argue that the embedded visual depiction of the Jabberwock (multiplied via the mirror-written *picture* poem and Tenniel's original drawing) function by means of transmedial addendums, visual translations of verbal nonsense, meant to assist the child reader in a nondidactic, ludic manner to face the interpretive rite of passage involved in the attempt to tame textual monstrosity. Finally, I will share some thoughts on the cultural transposition, linguistic transfer, and creative individual solutions emerging in Hungarian verbal and visual translations of "Jabberwocky."

The Implied Reader as Translator

The poem entitled "Jabberwocky" is a memorable specimen of image-textual monstrosity, a hybrid embodiment of verbal and visual nonsense haunting beyond the realm of the imaginable and the speakable. In

Carroll's novel, Alice first encounters the "Jabberwocky" nonsense poem scripted in a mirror-writing in the inverted world behind the Looking-Glass. She initially decodes it as a strange picture language she cannot read since the words turned topsy-turvy, from right to left, are perceived by her more as nonfigurative images than linguistic signs. For Alice, the Kcowrebbaj/Jabberwock, whom she never actually meets, remains a thingless name, a referentless signifier, a grotesque (image)textual[1] product that never gains a fleshly physical embodiment. Nevertheless, its overarching presence invades the fantastic diegetic universe like the ghostly aura of its absent author.

Accordingly, what fascinates and frustrates Alice is not so much the in/vincibility of the unseen mythical predator but the overall in/comprehensibility of the nonsensical discourse about it. Starting with the emblematic sentence, "Twas brillig, and the slithy toves / Did gyre and gimble in the wabe; / All mimsy were the borogoves, / And the mome raths outgrabe" (Carroll, *The Annotated Alice* 155)—the strange lines of the embedded poem evoke in Alice a reaction that mirrors the readerly response to the fictional frame-text in which she features as a protagonist.

> "It seems very pretty," [...] "but it's *rather* hard to understand!" [...] "Somehow it seems to fill my head with ideas—only I don't exactly know what they are! However, somebody killed something: that's clear, at any rate...." (156)

What is at stake here is the manageability and enjoyability of the textual monstrosity of nonsensical representation testing the limits of imaginability.

Alice is put on trial as a translator, and as Björn Sundmark highlights, her reactions to "Jabberwocky" (it sounds vague, but she believes to get the gist of it, plus it seems pretty) resonate perfectly well with novice language learners' experience of "just having enough knowledge to understand what a text is about, even when missing every other word. And what is lacking in comprehension is made up by the experience of novelty and fascination that the new language carries with it" (Sundmark, "Some Uffish Thoughts" 44).

Although no ultimate meaning is assigned to the "Jabberwocky" poem, Alice still functions as a clever reader/translator, maturing into a creative artist in her own right and managing to find her way in the wondrous land of (non)sense-formation. First, as a true bibliophile, right after her arrival to the other side of the mirror, she spots the book and gives a try on reading the poem. Second, a bright thought strikes her to hold the lyrics in the glass and succeeds in "making the words go the right way again" (Carroll, *The Annotated Alice* 154). Third, she easily recalls word-by-word the seemingly gibberish poetic text to Humpty Dumpty. Fourth, she takes an active part in finding possible explanations for its semantic incongruities and, while verbally toying with the Eggman, "makes thoughtful remarks" "surprised of her own ingenuity" (226). Finally, she light-heartedly leaves meanings unresolved, open-ended, and polysemic on both of her interpretive attempts of the Jabberwocky poem: at the beginning when she gracefully floats out of the Looking Glass House, curious of the adventures awaiting her, and later when she interrupts Humpty's despotic attempts at linguistic mastery, challenging him further on by mentioning another undecodable poem authored by Tweedledee.

As Sundmark points out, the Jabberwocky poem—a singularly written text amidst a plethora of orally presented rhymes and songs—is cleverly *translated* by Alice from one medium to another, moving from one cognitive register to another: (1) She converts the reversed sign system into a readable version; (2) she recodes the written text through its memorization; (3) she shifts her mental imagetext of the poem into the oral medium; (4) she has Humpty Dumpty recode the text into the "mock-philological register"; and (5) she invites readers' (re)interpretations of these "textually represented transactions," which inspire all "to read imaginatively!" (Sundmark, "Some Uffish Thoughts" 182).[2] Hence, Alice matures into *translator* as creative (co)author.

Illustration as Translation

A picturebook could roughly be defined as the iconotextual unit of words and images interacting with one another in a multimedial, polyphonic artwork that—by fusing a variety of semiotic modalities, verbal and visual devices, codes, and styles to strategically "push at the borders of convention" (Nodelman 69–80)—may eventually speak to a larger readership than its traditionally intended target audience of young children. As Riitta Oittinen points out, it is extremely challenging (and perhaps unproductive) to try to differentiate between an "illustrated story book" and a "true picturebook." The former, by its generic definition, is meant to tell the story with words amplified by pictures that are nevertheless dispensable in the comprehension of the story; the latter, traditionally aimed at pre-readers, tells a story entirely in pictures and "says in words only what pictures cannot show" (Shulevitz in Oittinen and Davies 5).

Lewis Carroll's Alice tales are perfect examples of the difficulty of this distinction. The author's own rudimentary sketches in the first version of *Alice's Adventures Underground* (presented as a giftbook to the muse, Alice Liddell, on the Christmas of 1864) and then his close collaboration with *Punch* magazine's cartoonist John Tenniel (whom he hired to illustrate the 1865 *Wonderland* and the 1872 *Looking Glass* editions published with Macmillan) suggest that Carroll found it important to tell his stories in both words *and* images. However, this iconotextual dynamics, so crucial to the Carrollian fantasy realm, is not a matter of the age of the target audience. *Nursery Alice* (1890), a shortened version Carroll created for pre-reader children "from nought to five," includes only half of the unabridged edition's original Tenniel drawings—albeit in an enlarged, colored format and with plenty of interpolations from the narrator. By pointing out details in the pictures and asking questions to enhance audience interaction, these asides oddly conjoin the evocation of an oral storytelling atmosphere with the foregrounding of visual details. Hence, *Nursery Alice* can easily be labeled as an oral–pictorial "distillation" (Susina) of Carroll's original unabridged novel.

The vital narrative engine of the Alice tales is commonly associated with language games, which are enjoyable without the illustrations—in a radioplay, an audiobook, or Project Gutenberg digital library's online text-only format. Yet the illustrations certainly add an extra special flavor to the text. In some instances, the images seem straightforwardly inevitable when, as if words failed a trustworthy representation of a fictional reality, the author points out of the text directing his audience toward the illustration: "If you don't know what a gryphon is, look at the image" (Carroll, *The Annotated Alice*). In this case, it is the illustrator's responsibility to guide the reader–spectator's mental imaging of the mythical beast, to translate the unspeakable into a pictorial form. This complementary visual addendum (the image of the gryphon) should make an integral part of the verbal translation of the text, too—bearing in mind that translation involves rereading and rewriting (Oittinen 1), as well as *reimag(in)ing*. Conforming with major criteria of harmonizing the textual and visual translation of the Carroll–Tenniel text-image, illustrations should enhance the nonsensical confusion of text by depicting signifiers which lack clearly identifiable signifieds.

The ambiguity of the mock heroic ballad's monstrous Jabberwock prevails in a visual form in John Tenniel's drawing that remains one of a kind, a still decisive depiction we have of the enigmatic monster that earns the status of metafictional imagetext. Readers' imagination is teased by the literary nonsensical ambiguity of this "whiffling," "burbling" creature with "jaws that bite" and "claws that clash." What has been in the poetic text a name without a thing[3]—a referentless signifier, an indefinable, faceless, formless enemy—keeps its ambiguity in pictorial form, too. Tenniel pays visual homage to the literary text's potential inspirations of legendary, mythological, and scientific origins. His Jabberwock is a degenerate mutant, a dead-end on the Darwinian evolutionary scale, a grotesque parody of monstrosity, a chimeric hybrid just like nonsense discourse's portmanteau word coinages: "two meanings packed up into one word" (*The Annotated Alice* 27). It fuses features of a dragon, a griffin, a dinosaur, an insect, and the legendary Lambton Worm, all dressed in a petit-bourgeois checkered waistcoat. In contrast with Elizabeth Sewell's arguments on illustrations' "anchoring potential," the visual translation of verbal ambiguity does not delimit meanings. It rather provides further stimuli for readers' mental imaging, allowing it to prosper simultaneously in multiple directions, entertaining then rejecting fantasy fragments, giving various temporary, kaleidoscopic physical forms to the ideas evoked by referentless verbal references. Notably, the Jabberwock was initially planned as a frontispiece to *Looking Glass*, but it became relegated to the middle of Chapter One to calm the mothers of prospective readers who feared that this "too terrible monster" would "alarm nervous and imaginative children['s]" fantasies (Gardner in Carroll, *The Annotated Alice* 163).

Charged with metanarrative (metapictorial) implications, the picture of the frail childlike knight ready to slay the monster with the vorpal sword can easily represent Alice, the implied reader struggling with textual monstrosity—the impossible challenge of making sense of nonsense. Michael Hancher also highlights that the "beamish boy" knight closely resembles Alice on another Tenniel illustration portrayed in a similar posture: from the back, with her long hair hanging down, and slightly leaning back to look upon another nonsensical Wonderland figure, Humpty Dumpty. The Eggman takes the Jabberwock's place in the image-composition while he explains word-by-word the mirror poem about the monster to Alice; he takes the maximal interpretive liberty, defines meanings arbitrarily at whim, and "makes words mean" "just what [he] choose[s] them to mean—neither more, nor less" (Carroll, *The Annotated Alice* 224). Thus, in my view, Humpty Dumpty simultaneously tames monstrosity by framing it (monstrous meaninglessness) into meaning and becomes monstrous himself by tyrannically usurping himself the right to make sense, to make words mean whatever he wants them to mean. While Hancher claims that this scene is "an androgynous projection of Alice's fears" (75), I would go further along these lines and stress, yet again, the imagetextual dynamics of Carroll's text.

The exchangeability of the monstrous illustration of the Jabberwock, too terrible to imagine and terribly unimaginable, and of his mis/interpreter Humpty Dumpty—whose absurd explanation's vain verbalization assumes to transform via a medial shift unimaginability into speakability but eventually cannot produce but an incomprehensible, unmasterable, unspeakable code (of communication)—further increases meaninglessness instead of resolving it. Accordingly, the fundamental fears of the Unimaginable and the Unspeakable, of visual and verbal representational dead-ends or thresholds of sense, like anxieties concerning the tyrannical control over meaning (represented by Humpty Dumpty) and the excitement about the anarchic proliferation of meaninglessness (represented by Jabberwock), go hand in hand. They constitute the major adventure and adversaries for the implied reader/spectator/translator Alice. This can only be eased by consolatory nonsense equally functioning on verbal and visual planes, resisting the solidification of signification.

In fact, "Jabberwocky" is a phantom imagetext haunting the Carrollian oeuvre. In 1855, Carroll wrote a pastiche called "A Stanza of Anglo-Saxon Poetry" for the amusement of his brothers and sisters.[4] It was this stanza he recycled twenty years later with some minor modifications as the first quatrain of "Jabberwocky." He later complemented the poem with a glossary explaining the difficult words that, funnily enough, on several points deviates from Humpty Dumpty's interpretations. This paratextual commentary addendum in search of the unattainable "real meaning" has been expanded ever since—most famously by Michael Gardner's notes in *Annotated Alice*, Peter Heath's speculations in *The Philosopher's Alice*, and more recently by Mark Burstein's updates in Norton's *150th Anniversary Deluxe Edition.* While the Jabberwock was explicitly mentioned (along with further invented words/creatures as the bandersnatch and the jubjub bird) in the text of Carroll's succeeding 1876 mock heroic ballad, the grotesque sea quest story, *The Hunting of the Snark*, I believe that a visual prefiguration of the character already lurks on the pages of the Wonderland adventures preceding the journey through the Looking Glass.

As a precursor of the pair of Tenniel illustrations I analyzed above (of the Alice-double beamish boy knight fighting the Jabberwock and of the iconographer Alice consulting the eggman Humpty Dumpty on the meaning of the nonsensical poem about the Jabberwock), in Wonderland we find an image of Alice portrayed from the back (recognizable because of her wavy hair and striped stockings) lost in a discussion with yet another paradigmatically nonsensical being, the Cheshire Cat, who spells out the major guideline to the fantastic dream realm: "we are all mad here." This axiom suggests that the only rule of making (non)sense here is that there are no rules—language ceases to function as a convention-based sign system—yet Alice and the readers must turn into co-producers of nonsense. Moreover, verbal nonsensification is visually enhanced by virtue of the ingenious book-design characteristic of the Carroll–Tenniel cooperation: the reader's single flip of the page back and forth induces an optical illusion whereby the Cheshire Cat is made to disappear with only its grin remaining behind (on the page beneath the one where it is shown in its full bodily integrity), whereas Alice contemplating the cat becomes lost amidst the text and disappears, "overwritten by words" beneath the grin (Wong 146). Representation literally vanishes as image dissolves into text, text fades into meaninglessness, and the only thing that remains behind as a surplus of signification is the grin generated by the self-deconstructing imagetext. A grin shared by the Cat, Alice, and the Reader alike.

It is difficult to tell which of Nikolajeva and Scott's categories (249–259) apply to the word–picture dynamic here, balancing between symmetrical, enhancing, complementary, counterpointing, and contradictory. Carroll and Tenniel's imagetextual hybrid invites "voyure"—a mode of interpretation that iconographer Liliane Louvel coined by mixing two words, contemplative *la lecture* (reading) and transgressive voyeuristic *la vision* (seeing), in a portmanteau that seems fitting for the study of literary nonsense filled with language games *and* visual puns alike (109). In the *Alice* books, the interaction between word and image enhances the fluidity or even *liquidification* of the signifying process by setting up, in Hillis Miller's words, an "oscillation or shimmering of meaning in which neither element is prior to the other, since the pictures are about the text, and the text is about the pictures" (in Hancher 113). Interpretations will vary depending on the different meanings added to the text—by Carroll's original illustrations, by Tenniel's line drawings (which eventually eclipsed Carroll's to become canonical, and which were updated from one volume to the other to make the heroine's dress contemporaneous with the readers' [Nières 198]), and by all future artists' revisionings of Wonderland—which all contribute to the transmedial extension of the storyworld.

Hungarian Translations

The illustrations' allusive instead of denotative quality is further enhanced when a foreign audience reads the text in translation and project their own culturally specific significations on the image, regardless of the translation adapting Lawrence Venuti's strategy of domestication or foreignization. The most obvious cultural interference for Hungarian readers facing Tenniel's portrayal of the Jabberwock comes from the Hungarian folk tale tradition's popular figure, the dragon (or dragonlizard) (*sárkány/gyík*). This folkloric creature, an intermediating character connecting the supernatural and the material realms, mostly appears as a winged reptilian or serpentine beast fusing interspecies characteristics. Often, it is believed to metamorphose into its new form from an old pike or a rooster unseen by human eyes for seven years. It mingles its bestiality with anthropomorphic features: It can master human speech; has human emotions and ambitions; loves wine, song, dance, and human maidens; and bears human facial features on its seven heads. But it can also be tamed and used as a mount by the trickster Garabonciás. It can rise from a whirlwind and alternately represents intellectual chaos or ultimate wisdom, while its slaying signals a coming of age initiation rite. As an ultimately scary ambiguous figure dwelling in the Hungarian collective unconscious, it resonates perfectly with the Jabberwock as the embodiment of the Unimaginable itself.[5] Interestingly, however, none of the literary translations acted explicitly on this allusion. It is only Tim Burton's cinematic adaptation of *Alice*—which excelled in a hyperrealistic realization of "what has never been" by virtue of a cutting-edge 3D CGI technology bringing life-like fantastic beasts onto the silver screen—that embraced it in its Hungarian dubbed version by calling the Jabberwock alternately "Dragonlizard" (*Sárkánygyík*) or with a pun "GuileLizard" (*Ármánygyík*) (*sárkány*, meaning dragon, rhymes with *ármány*, guile). Thereby, the translation in dubbing neglected the original language game in favor of "tweaking the source text to reflect the sense of humor and culture of the target country" (see Crookes) with this easily accessible Hungarian folktale reference.

As Riitta Oittinen explains, "a picturebook is an art form with many different voices to be heard and seen," including the voices of the author, the illustrator, the translator, the publisher, and the different readers, children and adults (Oittinen and Davies 5). Each is influenced by their own selves and personal prehistories as well as the norms, poetics, and ideologies prevailing in their sociocultural milieu. Every time a book is translated, "it takes on a new language, a new culture, and new target-language readers" in this complex Bakhtinian multivoiced, mutifocal situation, where all the agents of meaning formation listed above meet and influence one another (Oittinnen 1). In the following, I wish to show how Carroll's nonsensical textual monstrosity is multiplied in many forms throughout its different Hungarian translations.

Although *Alice's Adventures in Wonderland* (1865) has six different Hungarian translations (Altay 1927, Juhász 1929, Kosztolányi 1935, Szobotka 1958, Varró 2009, and Szilágyi 2013), its sequel *Through the Looking Glass* (1871) has so far only been translated twice into Hungarian (Révbíró and Tótfalusi, 1980, Varró and Varró, 2009). Both *Looking Glass* translations result from a creative cooperation: a different author took charge of the prose narrative and another of the lyrical inserts. In the 1980 edition, translator Tamás Révbíró[6] was assisted by poet István Tótfalusi (who also translated Charles Perrault, A. A. Milne, J. M. Barrie, and Astrid Lindgren). The text was enhanced by Tamás Szecskó's illustrations (with no image of the Jabberwock). The 2009 joint translation of *Wonderland* and *Looking Glass*, decorated by Tenniel's original illustrations, is a collaboration of the Varró siblings—children's writer Zsuzsa Varró and her brother, perhaps today's most popular Hungarian children's poet, Dániel Varró. This is the only edition that features the "Jabberwocky" text accompanied by an illustration, hence the only complete rendition of Carroll–Tenniel imagetext. However, the novel's single most well-known embedded nonsense poem challenged further attempts at textual transposition. Two more Hungarian Jabberwockies occupy the high and low end of the cultural spectrum, respectively. One was published in the 1970s by a prestigious publishing house in a poetry volume (targeting adult audiences) and authored by Sándor Weöres, a veritable master of language games who has been acknowledged as one of the greatest twentieth-century poets of the Hungarian literary canon and nominated several times for the Nobel prize. The other was uploaded online in 2011 on a website for amateur poets by a young blogger named Balázs Zs Jónai, who also shared on the internet his Hungarian translations of *The Hunting of the Snark* and the omitted chapter of *Looking Glass* ("Wasp in a Wig") along with several Carroll poems.

When it comes to translating the sound of nonsense, all four of the Hungarian translations respect the original poem's formal, structural outlines. They keep the strong rhythm, the iambic meter, the ABAB rhyme scheme, the atmospheric setting frame of the repeated first and last stanza, the conventional ballad form, the simple syntax, and the straightforward storyline (including the retrospective account of the preparative warning, the heroic quest, the defeat of the monster, and the laudation of victory).

The Carrollian language games, neologisms, and portmanteaux are easily integrated within Hungarian translation because of the extremely metaphorical, poetic nature of Hungarian language, which abounds in word coinages due to the nineteenth-century Language Reform. Throughout this Language Renewal (*nyelvújítás*)—carried out by leading scholars, authors, poets, translators, linguists, and neologists—thousands of words were created or revived so as to enable the Hungarian language to clearly communicate scientific, legal, and literary notions that had been previously expressed in Latin or German (the official languages of the Habsburg-controlled Austrian empire that the Kingdom of Hungary had been part of since 1526, until the establishment of the dualist state of the Austro-Hungarian Monarchy in 1867). Due to these politically committed creative efforts to regenerate Hungarian language and literature, Hungarian became an official language of the nation. Its vocabulary was enriched by regional dialectical expressions spread nationally, archaic extinct terms revived, domesticated phonetic transcriptions of foreign words, and an amazing range of new words forged by derivative suffixes, the portmanteau blending of multiple words' sounds and meanings, onomatopoeic or synesthesiac word coinages (see Laakso). While some of the too-complicated neologisms disappeared, many creations of the Linguistic Reform's lexical expansion function today as fossilized metaphors, the poeticity of which is nearly imperceptible for native speakers of Hungarian. This language's historical background is a reason why the nonsensical neologisms of "Jabberwocky" translate well into

Hungarian. (This is a language that calls a sibling a *testvér*, literally meaning a person who comes from the same "body and blood"; refers to a bird's beak with the mingling of the words "tube" and "nose," *cső* + *orr* = *csőr*; has playfully melodic terms for simulated anger and gentle scolding, *ejnye bejnye* and *irgum burgum*; and uses an expression that translates as "golden bridge" (*aranyhíd*) to describe the reflection of the sunset on the surface of a pond.)

As Pilar Orero, the author of a 2007 book on Spanish translations of "Jabberwocky," has pointed out, the real challenges of translating nonsense dwell not in the syntactic but in the semantic dimensions. I would elaborate on her claim by arguing for the significance of the challenge dwelling in the interplay of the semantic and the phonetic dimensions. The latter involves the very sensorial experience of linguistic performance that is foregrounded by literary nonsense, a genre where sounds precede sense.

I agree with Kit Kelen that the ideas of nonsense, topsey-turvydom, and misrule well established in Victorian England for the poetic entertainment of children peak in Carroll in irony, "a trope of multiplicity, involving an oscillation yet simultaneous perception of plural and different meanings" (77). Alice is uncertain what sense to make of "Jabberwocky" because the dramatic plot ("somebody killed something") and the heroic ballad genre (i.e., the sense and the form) stand in contrast with the vocal play, the sounds of the odd and funny noises and voices permeating the poem. As Sundmark suggests, a major question of "Jabberwocky" (that may be the most difficult to convey in translation) is how the mock heroic register is played out: "the poem is both heroic *and* parodic at one and the same time; the tension is never wholly resolved [and] this ambiguity [must] resonate in the translations" ("Some Uffish Thoughts" 53). The heroic feat, the drama of the traumatic encounter with the monster, is counterbalanced by the extreme oral delights of the text (the corporeal physical pleasures of producing sounds, as this is a text meant to be read/performed out loud), which simultaneously prevent words from making/fixing meanings and foster the proliferation of significations.

All the Hungarian translations provide target-language equivalents of the trademark Carrollian language games, using invented words, portmanteaus (blending the sounds and merging the meanings of two words), alliterations, onomatopoeia, and other forms of the vocal play with sounds preceding and predominating the sense conforming to the logic of the genre. The meaning-troubling, phonetic, auditory experiments are so vital in the text that some argue the name of Carroll's monster can be decoded by etymologically tracing it back to old Scottish words "jabber" standing for "babble" and "wocky" meaning "voice." An *Introduction to Poetry* study guide for students goes as far as to translate "Jabberwocky" along the line of a mock didactic message as "Be careful of the Babble Voice my son, [...] stay away from the fuming and furious creature that robs sentences of their meaning!" (Cengage Learning). This interpretation paraphrases a dictionary definition of literary nonsense: sounds precede and overwhelm sense, or in the Wonderland Gryphon's words, adventures predominate explanations. The extremely acoustic nature of the text is further stressed by the gaping mouth of the Jabberwock on Tenniel's illustration, which offers an odd visual representation of a silent scream, a voiceless cry that opens up transmedial, audiovisual dimensions throughout the interpretive experience.

Tellingly, the active, kinetic verbs of action describing the wrongdoings of the Jabberwock have to do with sounds ("Came *whiffling* through the tulgey wood / And *burbled* as it came!"). These sounds are preserved in the Hungarian translations. In Tótfalusi's version, onomatopoeias are turned into neologisms referring to the sounds of flapping wings and snoring-muttering noises which accompany the advent of the monster ("hussongva és mortyogva jött"). In Weöres' version, "it is foaming/gurgling while shaking" as if arriving on water ("bugyborékolva ráng"). In Varró's, it is "panting/moaning and clomping" ("bihálva csörtetett elő"), the sound of its heavy tread/footsteps complementing its strange vocal performance. In Jónai's, it simply "arrives grunting" ("morgva érkezett meg"). Hence, the Jabberwocky takes many forms throughout its Hungarian translations; it can be a creature of air, of land, and of water alike. Moreover, the strange sounds it makes once again resonate well with the Hungarian folkloric dragon that can alternately use human language, emit an animalistic howl, or simply spit fire by means of brutal nonverbal communication.

The acoustic qualities of the text are maximized in the first translation of *Through the Looking Glass*. In its "Jabberwocky" version, the translator, Tótfalusi, turns four of the five verbs of the first atmosphere-setting ballad stanza into onomatopoeic verbs, suggesting some kind of nonhuman vocal agency—evoking the cooing of doves, the purring of cats, the grunting of swine, and the munching of beasts ("turboltak, purrtak," "nyamlongott, bröftyent"). Yet Humpty Dumpty's explanations relate these noises to mechanical sounds of rotating turbines ("turbine"), spinning drills ("pergő fúró"), and roaring whistles ("bömbölés," "füttyentés," "röfögés"), enhancing the grotesque in-between quality of these uncategorizable auditory sense impressions. Through his choice of onomatopoeic verbs describing the monster's actions, the Eggman completely redefines the Jabberwock. His explanation transforms the chimeric composite of different beasts (the Jabberwock's original textual and pictorial portrait) into a machine–animal–human hybrid.[7] His rewording/reimagining supports the claim that "Humpty Dumpty is an unreliable translator, and his explanations produce more nonsense, as do all subsequent attempts at 'translating' the poem, whether to English (in learned annotations) or into other languages. Translation makes for the proliferation of nonsense" (Sundmark, "Some Uffish Thoughts" 46).

Perhaps it is this vocal, sonoric quality of nonsense that is meant to be encapsulated in the title choice of three of the four Hungarian translations, which all play puns on bird names. Birds might come into the picture for various reasons. Bird is the animal most automatically associated with the sound, the song, or the cry it makes. Bird is also a common metaphor for artistic creativity, poetic agency, and lyrical self's struggle for meaning construction (e.g., Keats' skylark, Shelley's nightingale, Baudelaire's albatross, Poe's raven, etc.). On the other hand, mythical bird figures also play a significant role in Hungarian folkloric imagery. Turul is a bird of prey with supernatural powers, a protective spirit of the dream realm, and a progenitor of Hungarian forefathers.

The first ever Hungarian translation of "Jabberwocky," published by Weöres in 1958, calls the creature "Szajkóhukky"—combining the word for jay/magpie (*szajkó*) with the phonetic transcript of hickups (*hukk*) and adding an "y" to the end, as if to indicate the archaic, obsolete, or alien nature of the word. The word "jay" (*szajkó*) is also used in Hungarian, transformed into a verb to refer to the speech acts of a person who is a chatterbox or keeps repeating the same thing all over again in an annoying way. Therefore, over-verbalization and under-verbalization, being a chatterbox or being silenced because of hickups, are condensed into one single name here.

The two other Hungarian bird translations of Jabberwock have been possibly affected by the first translator's inventive solution. Tótfalusi's "Gruffacsór" (1980) evokes the fantastic beast gryphon (*griff*) with a difference of a single letter (*gruff*), whereas the second part of the name evokes "beak" (*csőr*) and "snatch/steal" (*csór*) and even a miserable tramp's figure (*csóró*). Thus, a mythological creature—the gryphon, who rises from its ashes just like sense arises somehow from nonsense—is demythologized, demystified by lowly connotations in honor of the mockery in the mock heroic poem. In Jónai's 2011 version, "Vartarjú" plays a pun on the word "crow" (*varjú*) by breaking it up in two syllables and inserting the word "bald" or "barren" (*tar*) in between, investing the name with ominous implications augmented by the first syllable meaning "scar" (*var*).

Since Weöres and Jónai translate merely the individual poems, separated off from the context of the novel, they do not have to bother with Humpty Dumpty's explanations of the words' meanings and enjoy the most poetic liberty. Out of the four translations, these two are the most condensed, free, and perhaps most domesticated versions. Although one might wonder if "domesticated" is the right words since the expressions they use sound as nonexistent neologisms—unfamiliar, invented words for Hungarian ears, too.

The Weöres translation has the fewest unfamiliar words. It changes the syntax and eliminates the portmanteau of the first line to set the atmosphere with a classic fairy-tale beginning, evocative of the well-known phrase "Once upon a time." Its "in medias res" beginning "there was a brillős" ("Volt egy brillős") informs about the legendary existence of a (presumably shiny and noisy) fictive beast instead of denoting the time of day as in the original "Twas brillig" (decoded as "it was four a clock, time for broiling things for dinner"). The temporal reference is preserved but set on a mythical plane. Weöres and Jónai, who deal with the poem decontextualized from the prose narrative, also feature the name of the monster without a pronoun, thereby suggesting that it is a proper name of a one of a kind being, unique in its existence. In contrast, Tótfalusi's use of a definite pronoun suggests that the beast's name refers to a species instead of an individual.

The Varrós translation of the two Alice books was driven by the agenda to produce a Hungarian text that is more truthful to the original source-language narrative than its predecessors. It aimed to challenge both an extremely domesticated translation, entitled *Evie in Fairyland* (*Évike Tündérországban* by Hungarian poet Kosztolányi (1935), and its slightly updated, revised version by Szobotka (1958) that has been the most popular among Hungarian readers of Alice. Varró's title, in line with the fidelity criteria, is the only one that respects the Carrollian distinction between the name of the monster (the Jabberwock) and the title of the heroic ballad about the slaying of the monster (Jabberwocky). The sound of "Hergenyörciád" resonates with the title of Homer's epic poem *Iliad*, yet the word remains perfectly nonsensical, undecodable, unrelatable to any Hungarian notions. We are troubled by this sort of referentless signifying short circuit since the syntax clearly seems to suggest there is an external material referent, yet the words refuse to mean any concrete thing and (due to this discursive inventiveness) grant full "visibility" to the translator (in Venuti's sense). It "gives the impression of being in a different [foreign] language only partially understood" (Sundmark, "Some Uffish Thoughts" 55), conforming to the logic of the novel where Alice's first reaction to the poem is "It is all in a language I don't know." As Sundmark highlights, this effect of the foreignness of the text is "first created by the inversion of the text, then by the 'hard words,' and finally by Humpty Dumpty's translations and explanations" (54). "'Jabberwocky' requires translation even in the original, and thus never can be an 'invisible' translation (in Venuti's sense)" (54). Varró manages to retain this foreignness in translation by evoking in his title the Greek epic tradition, familiar in its unfamiliarity to the target-language populace since extracts from *Iliad* and the *Odyssey* are compulsory readings for Hungarian teens in the final class of elementary or the first year of middle schools.

This intertextual allusion also indicates a slight change in the target readership of this translation. Zsuzsa Varró confessed in interviews that the Alice books frightened her when she read them as a child. She recalls having found them "chaotic, dumb, and depressing" (Varró). However, the Varró siblings' Hungarian translations did not soften or infantilize the source text but reproduced it with a slightly older, young adult audience in mind. The Varró translations still talk about/to child readers, but the translation of the polysemic puns in the Mock Turtle's lessons, for example, contains word pairs that no child reader would likely be able to understand. Carroll's original homonymous puns—toying with similarly sounding but funnily distorted terms, mocking mandatory subjects of the school curriculum—"reeling and writhing" instead of "reading" and "writing," along with classes called "uglification, derision, and distraction," are easily accessible to younger audiences. Yet this joke remains difficult to grasp for the child reader in the translation by the Varrós, who use loan words of Latin origin for both the name of the original school subject and its parodic revision. "Derogate" is meant to be the funny Wonderland version of "derivate," but the humor incited by the tension between the familiar and the unfamiliar is lost when both the right and wrong versions are similarly unknown to the child reader/listener. (The word pairs are "imponálás"/"integrálás," "derogálás"/"deriválás," and "spriccelés"/"skiccelés," see Kappanyos 215.)

The shift in the target audience in the Varró translation is an example of how the translator's individual notion of childhood and the child reader might considerably affect the textual construction of the target-language end product. The Alice brought to life in the Varrós' Hungarian translation clearly differs from the naïve, innocent, somewhat silly Evie (Évike) of Kosztolányi's 1935 domesticated Hungarian translation and even from Carroll's own idealized Dreamchild. Her curiosity remains the same, but she appears perhaps more world-wise, even rebellious. She dares to ask questions and, in a way, seems to come of age, to gain self-confidence in a *Bildungsroman*-like manner throughout the progress of her adventures (see Kappanyos). The Varró-translation's physical form also attests a more mature target audience: This edition published the two volumes in one single large-format book, with a heavy, hardcover, coffee-table-book-like edition that is undoubtedly too heavy to hold or to leaf through for small hands. It is exciting to note that the Varró translation was published in the same year as Tim Burton's 3D CGI Wonderland film adaptation (2009), and it resonated well with the film's strategic address of a dual/crossover audience and the portrayal of Alice as a Jeanne d'Arc-like action heroine who rebels against repressive social mores to liberate Underland from the tyrannic rule of an evil monarch (see Kérchy, *Alice in Transmedia Wonderland*). Theirs is an agile Alice for the postmillennial era.

Imaging Alice as a preteen, co-productive agent who is diligent in the "making sense of nonsense" also features on an earlier Hungarian pictorial translation of Carroll's classic. In the 1980 Hungarian *Looking Glass* edition's illustration, heavily influenced by Tenniel's etching, Tamás Szecskó's pencil drawing portrays Alice struggling with textual monstrosity. However, the sword in the hand of the beamish boy fighting the Jabberwock is replaced by a pencil in the young girl's hand, as she is taking notes of Humpty Dumpty's explanations of the apparently meaningless poem. Hence, reader Alice matures into a co-authorial figure and can cope with nonsense more efficiently. An inquisitive Alice who is relentlessly asking questions about meanings, while being aware of the absence of ultimate answers, highlights the dialogic nature of the open text (in the post-modern narratological sense of the terms). This image of a note-taking Alice—in conversation with an incomprehensible creature who speaks a language (and a logic) radically different from her own—stages the interpreter's curiosity and creative agency and also perfectly encapsulates the ambiguous features of nonsense fairy-tale fantasies explored on a variety of levels by current Carroll criticism. Alice's endless questions keep her in the domain of childhood but also lead her toward a knowledge that will allow her to become an adult, as Kelen suggests (77). Her Platonic conversational strategy proves both her narcissistic yearning for self-knowledge and her solidarious worldview and empathic relationality, which Gillian Beer called the major narrative engine of *Wonderland*. Alice's conversational questioning accompanied by note-taking illustrates the odd mixture of oral and written registers, explored by Sundmark's 1999 monograph and singularly captured on Szecskó's illustration. This illustration, by offering a visual translation of linguistic agency, holds mirror to *Looking Glass*'s fundamentally ekphrastic text, grounded in verbal descriptions of visually intense dream scenes.

These contradictory, complementary aspects—text and image, sound and sense, original and its many translations—interact in a variety of complex ways to create an exciting, polysemic, multimodal Wonderland experience. They facilitate an "additive comprehension" characteristic of the Jenkinsian "transmedia extensions." Further Hungarian variations on the Jabberwocky theme—including physical theater troupe Szárnyak Színháza's exploitation of the text's kinetic potential via modern ballet's defamiliarization of movement, the grotesque imagery of Geraldine Uzoni's black and white stop-motion animation, or Barbara Palvin's "soft storytelling" whisper performance—prove that there is practically no limit to enhancing the immersiveness of a storyworld by adding innovative perspectives that each invite us to repeatedly revise our understanding of an old/renewed fictional reality.

Notes

1. Furthermore, as Mou-Lan Wong convincingly suggests, in between the two (reversed and regular) variants of the typographically distinguished (indented, italicized) "Jabberwocky" poem, we find a primary text segment (in which Alice ponders about the proper decoding of a Looking glass book) that functions as "a verbal looking glass that alters the visual outlook of the two variations" (141), and also, I might add, as a separating, interconnecting, self-reflexive plane that separates, interconnects, and *translates* mirrored image(text) into nonsensical text(image). Given that the mirrored

verse is produced by a method mainly reserved for illustrations, "the text not only interacts but also actually transforms the visualization of the verse-image and reconfigures it as acceptable verbal text" (Wong 141).

2. Sundmark adds that "In oral delivery, the realization and interpretation of narratives are jointly produced by teller and audience, whereas written texts bear witness to the physical separation of narrator and addressee—the sheet of paper (or looking glass) which, metaphorically speaking, separates author and reader and necessitates a recoding from one medium to another" (182–183).
3. If the "thing without a name" is a common trope in Gothic/horror fiction causing an uncanny dread of *je-ne-sais-quoi*, the "name without a thing" provokes a just as much chilling cognitive dissonance by the transgression of linguistic codes of representation. Rosemary Jackson sees this gap between the signifier and the signified a major characteristic of the modern fantastic, a disjunction of word and object-world inducing a "signifying short circuit" she derivates from Beckett's *Molloy*: "There could be no things but nameless things, no names but thingless names" (qtd. in Jackson 38).
4. Roger Lancelyn Green has pointed out that "Jabberwocky" could have been loosely inspired by Friedrich de la Motte Fouqué's *The Shepherd of the Giant Mountains*, which was translated into English by Lewis Carroll's niece Menella Bute Smedley in 1846 (Gardner 195).
5. The image of the slaying of the Jabberwocky was replaced by Tenniel's less frightening image of Alice and the White Knight, while the cover of *Nursery Alice* features Gertrude Thomson's idyllic image of Alice fallen asleep by the side of her book from which the fantastic creatures made up in her dreams emanate to gently surround her.
6. His ars poetica as a translator was encapsulated in Ákos Fodor's haiku-poem Axiom: "You should try and help/ everything to be the way/ it is anyway") ("segíts mindennek/ olyannak lennie, mint/ amilyen úgyis").
7. Interestingly, the association of a chimeric Wonderland beast with a machine reoccurs in a "visual translation," on Tamás Szecskó's 1958 Hungarian illustration of the Cheshire Cat translated as a "wooden dog" (*fakutya*) in Kosztolányi's domestication. The "wooden dog" (*fakutya*) is a literalized embodiment of the Hungarian metaphorical idiom with an uncertain etymological origin: "to giggle/grin like a wooden dog," where the wooden dog stands for an archaic term for a sledge making a screeching sound like laughter on ice, or a boot-horn shaped like a smile. The word choice of Juhász's 1929 translation, "wooden puss" (*facica*) used the same pun in a less recognizable, more twisted form that nevertheless complied with Tenniel's illustrations decorating the 1929 Hungarian edition. Tamás Szecskó's illustrations depicted a grinning dog in two remarkably different styles: his first cutesy, mannered, light, sketch-like illustrations in the 1958

edition featured an obviously manufactured wooden dog, of the Hungarian vizsla breed, with bolts at its joints, benignly smiling at Alice portrayed as a rather prim little muss, while Szecskó's more refined illustrations to the 1974 and all the succeeding Kosztolányi-Szobotka translations included the picture of a sharp-toothed, grinning creature with a British bulldog's head and a massive wooden body, reminiscent of the Trojan Horse (see Kérchy, "Essay on the Hungarian Translations" 297).

Works Cited

Beer, Gillian. *Alice in Space. The Sideways Victorian World of Lewis Carroll*. Chicago University Press, 2016.

Burton, Tim, director. *Alice in Wonderland*. Walt Disney Pictures, 2010.

Carroll, Lewis. *Alice a Csodák országában*. Translated by Margit Altay. Budapest: Pallas, 1927.

———. *Alisz kalandjai Csodaországban*. Translated by Andor Juhász. Budapest: Béta Irodalmi Részvénytársaság, 1929.

———. *Évike Tündérországban*. Translated by Dezső Kosztolányi. Budapest: Gergely R., 1935.

———. *Alice Csodaországban*. Translated by Dezső Kosztolányi and Tibor Szobotka. Budapest: Móra, 1958.

———. *Alice Tükörországban*. Translated by Tamás Révbíró and István Tótfalusi. Budapest: Móra, 1980.

———. *Aliz kalandjai Csodaországban és a tükör másik oldalán*. Translated by Zsuzsa and Dániel Varró. Budapest: Sziget, 2009.

———. *Nursery Alice*. Macmillan Children's Books, 2010.

———. *Aliz Csodaországban*. Translated by Anikó Szilágyi. Cathair na Mart: Everytype, 2013.

———. *The Annotated Alice. 150th Anniversary Deluxe Edition*, edited by Martin Gardner and Mark Burstein. Norton, 2015.

Cengage Learning. *Poetry for Students, A Study Guide for Lewis Carroll's "Jabberwocky"*. Gale Group, 2001.

Crookes, David. "How to Dub a Film." *The Independent*. October 4, 2011.

Goldthwaite, John. "The Un-writing of Alice in Wonderland." *The Natural History of Make-Believe: A Guide to the Principle Works of Britain, Europe, and America*. Oxford University Press, 1996.

Hancher, Michael. *The Tenniel Illustrations to the "Alice" Books*. Ohio State University Press, 1985.

Heyman, Michael. "Introduction to Nonsense." *Bookbird: Journal of International Children's Literature*, vol. 5, no. 3, 2015, pp. 5–8.

Jackson, Rosemary. *Fantasy: The Literature of Subversion*. Routledge, 1981.

Jenkins, Henry. "Transmedia Storytelling." *Confessions of an Aca-Fan. The Official Weblog of Henry Jenkins*, March 22, 2007, http://henryjenkins.org/2007/03/transmedia_storytelling_101.html.

Jónai, Zs Balázs. "Vartarjú" *Napvilág Íróklub*, 2012. http://iroklub.napvilag.net/iras/39957.

Kappanyos, András. Bajuszbögre, lefordítatlan: Műfordítás, adaptáció, kulturális transzfer, Budapest, Balassi, 2015.

Kelen, Kit. "Where the Child Is Father. Republics, Expulsion, and the Rules of Poetry. Exploring Lewis Carroll's 'Jabberwocky.'" *Child Autonomy and Child Governance in Children's Literature: Where Children Rule*, edited by Kit Kelen and Björn Sundmark. Routledge, 2016, pp. 66–84.

Kérchy, Anna. *Alice in Transmedia Wonderland. Curiouser and Curiouser New Forms of a Children's Classic.* Jefferson: McFarland, 2016.

———. "Essay on the Hungarian Translations of Wonderland." *Alice in a World of Wonderlands: Translations of Lewis Carroll's Masterpiece*, edited by Lindseth, Jon and Markus Lang. Oak Knoll Press, 2015, pp. 294–299.

Kristeva, Julia. *Revolution in Poetic Language*. Translated by Margaret Waller. Columbia University Press, 1984.

Laakso, Johanna. "Hungarian Is No Idioma Incomparable: The Hungarian Language Reform in European Comparison." *Hungarian Cultural Studies. Journal of the American Hungarian Educators Association*, vol. 7, 2014, pp. 320–336.

Lecercle, Jean Jacques. *Philosophy of Nonsense: The Intuitions of Victorian Nonsense Literature*. Routledge, 1994.

———. "Translate It, Translate It Not." *Translation Studies*, vol. 1, 2008, pp. 90–102.

Louvel, Liliane. *Poetics of the Iconotext*, edited by Karen Jacobs. Translated by Laurence Petit. Ashgate, 2011.

Mitchell, W. J. T. *Iconology: Image, Text, Ideology*. University of Chicago Press, 1986.

Nières, Isabel. "Tenniel: The Logic Behind His Interpretation of the Alice Books." *Semiotics and Linguistics in Alice's World*, edited by Rachel Fordyce and Carla Morello. Walter de Gruyter, 1994, pp. 194–209.

Nikolajeva, Maria and Carole Scott. *How Picturebooks Work*. Routledge, 2006.

Nodelman, Peter. *Understanding Children's Literature*. Routledge, 1999.

Oittinen, Riita. "Where the Wild Things Are: Translating Picturebooks." *Meta*, vol. 48, no. 1–2, 2003, pp. 128–141.

Oittinen, Riita and Maria Gonzalez Davies, editors. *Whose Story? Translating the Verbal and the Visual in Literature for Young Readers*. Cambridge Scholars, 2008.

Orero, Pilar. *The Problem of Translating 'Jabberwocky'. The Nonsense Literature of Lewis Carroll and Edward Lear and Their Spanish Translators*. Edwin Mellen Press, 2007.

Sewell, Elizabeth. *The Field of Nonsense*. Chatto and Windus, 1952.

Sundmark, Björn. Alice in the *Oral-Literary Continuum*. Lund Studies in English 97. Lund University Press, 1999.

———. "Some Uffish Thoughts on the Swedish Translations of *Jabberwocky*." *European Journal of Humour Research*, vol. 5, no. 3, 2017, pp. 43–56.

Susina, Jan. *The Place of Lewis Carroll in Children's Literature*. Routledge, 2010.

Varró, Dániel és Zsuzsa. "Egy testvérrel semmi baj sincs." *Könyves-blog*, 2010, http://konyves.blog.hu/2010/04/26/a_varro_testverek_a_konyvfesztivalon.

Venuti, Lawrence. *The Translator's Invisibility*. Routledge, 1995.

Weöres, Sándor. "Szajkóhukky." *A lélek idézése: Műfordítások*. Budapest: Európa, 1958.

Wong, Mou-Lan. "Generations of Re-generation. Re-Creating Wonderland through Text, Illustration, and the Reader's Hand." *Alice Beyond Wonderland. Essays for the Twenty-First Century*, edited by Cristopher Hollingsworth. University of Iowa Press, 2009, pp. 135–155.

Metapictorial Potentialities

Translated Book Covers as Peritextual Thresholds: Comparing Covers of Greek Translations to Covers of Source Texts

Petros Panaou and Tasoula Tsilimeni

Emer O'Sullivan observes that "Children's literature has transcended linguistic and cultural borders since books and magazines specifically intended for young readers were first produced on a significant scale in eighteenth-century Europe" (1). Translation and transfer are the vehicles on which stories have been traveling across linguistic and cultural borders, at the same instance crossing spatial borders between nations as well as time boundaries from one era to another.

Ever since the inception of children's literature as a separate body of texts, imported children's literature has had an important place in Greece. Writing about nineteenth-century children's literature in Greece, Ntelopoulos reports the translation of 293 books, by 114 authors, in 510 different editions (38). These numbers become even more remarkable when compared to the much smaller production of original Greek texts during the same period: 204 titles, 98 Greek authors, 330 editions (Ntelopoulos 38). Even though imported titles currently comprise about

P. Panaou (✉)
University of Georgia, Athens, GA, USA

T. Tsilimeni
University of Thessaly, Volos, Greece

A. Kérchy and B. Sundmark (eds.), *Translating and Transmediating Children's Literature*, Critical Approaches to Children's Literature,
https://doi.org/10.1007/978-3-030-52527-9_9

one-third of the Greek children's book market, they still have a strong presence. Greek scholars of children's literature purport that translations have always had a significant effect on Greek authors' themes, content, and styles (Anagnostopoulos), while Europe's important cultural influence on Greece was partly mediated through the European texts imported for Greek children (Ntelopoulos). Dimitrios Politis identifies modern and postmodern European influences on contemporary Greek children's literature, naming the French author Michel Tournier and the Italian Gianni Rodari as two important influences (54).

In spite of the prominent place of translated children's literature in Greece, comparative or translation studies are scarce. This paper—a shorter version of which was presented at the 2018 IBBY Congress in Athens—aims to contribute to the development of the field, through a "contact and transfer study," a study that focuses on the exchanges between literatures from different countries, languages, and cultures (O'Sullivan 2005). To be more specific, the paper compares the covers of sixty-eight Greek translations to the covers of their source texts. These are children's and young adult books that have been translated and imported to Greece by two major Greek children's literature publishers, Patakis and Psychogios, over the past ten years. As Anna Kerchy observes,

> Mediation—whether in the form of adaptation, translation, or remediation—allows for the reevaluation of a variety of notions ranging from authenticity, textuality, authorship, audience agency, age appropriateness, storytelling, or imaginativeness, while foregrounding the ideological interests, the educational and ethical responsibilities, and the semiological complexities involved in the trans(pos)ition process. (4)

Looking at how cover images, titles, designs, and other peritextual elements are translated, altered, adapted, omitted, added, or replaced, we draw conclusions about the transfer and reception of Greek translations for youth, as well as about trends, preferences, and sensitivities in Greek children's and YA literature in general.

Covers as Peritextual Thresholds

Paratextual elements are features associated with a text but peripheral to the narrative itself (Genette). Genette sees two categories of paratextual elements: the peritext and the epitext. Epitextual elements are in some distance from the text (not part of the book), and these can be anything from book advertisements to author interviews and book reviews. Covers—along with back covers, title pages, tables of contents, end papers, dedications, etc.—are peritextual elements that are in close proximity to the text itself (they are part of the book).

Along with the rest of the peritext and the epitext, covers are "what enables a text to become a book and to be offered as such to its readers and, more generally, to the public" (Genette 1). It is not a mere boundary between the text and the world:

> [T]he paratext is, rather, a threshold, or—a word Borges used apropos of a preface—a "vestibule" that offers the world at large the possibility of either stepping inside or turning back. It is an "undefined zone" between the inside and the outside, … an edge, or, as Philippe Lejeune put it, "a fringe of the printed text which in reality controls one's whole reading of the text." (Genette 2)

Genette continues to emphasize the important role of this zone as "a zone not only of transition but also of transaction: a privileged place of a pragmatics and a strategy, of an influence on the public, an influence that—whether well or poorly understood and achieved—is at the service of a better reception for the text and a more pertinent reading of it" (2). Book covers are an important peritextual element that is undoubtedly worthy of analysis (Serafini et al.; Sheahan).

We maintain that the covers of translated books are even more significant as transition and transaction zones because they influence the public/reader's reception of a "foreign" text. The cover mediates not only between the real world and the fictional world of the book but also between two different linguistic, cultural, and often geographical worlds. Comparing the translated covers to those of the source text, then, can provide insights regarding the specific strategies employed to influence this transition/transaction/reception. It can also provide insights regarding the target readers' preferences, tendencies, and expectations, at least as perceived by those who create the new covers (translators, book designers, editors, publishers, etc.).

The Multimodal Ensemble Analytical Instrument (MEAI)

Genette lists multiple items that can go on a cover, from the name or pseudonym of the author and the title of the work to the cover illustration, genre indication, name of the translator, information about the publisher, name of the series, number of "editions" or "thousands," and even facsimiles of the author's signature (24). He then points out that,

> Usually these localized verbal, numerical, or iconographic items of information are supplemented by more comprehensive ones pertaining to the style or design of the cover, characteristic of the publisher, the series, or a group of series. Simply the color of the paper chosen for the cover can strongly indicate a type of book. At the beginning of the twentieth century, yellow covers were synonymous with licentious French books. (24)

In our comparative analysis of source and target book covers, we paid attention to all of the elements listed by Genette. One early observation was that the translator's name almost never appears on the cover of the translation, an indication of the translator's low status in Greece; those who create the covers do not think that it matters to the public who the translator is.

In analyzing the covers, we also took into consideration Serafini, Kachorsky, and Goff's Multimodal Ensemble Analytical Instrument (MEAI), which is a tool designed specifically to guide critical multimodal analysis of book covers—even though it was not specifically designed for analyzing children's book covers. We paid more attention to the first category of considerations listed in the MEAI (Serafini et al. 113), which is called "Intramodal Considerations" and includes the following:

Compositional Inventory: Textual Elements

- Title: How is the title/subtitle presented? (color, size, position)
- Linguistics: What Verbs /Nouns are used in the title? What do these suggest?
- Fonts: What are the characteristics of the fonts used? (weight, coherence, color, serif /sans serif, expanded or condensed)
- Author: How are the names of authors/editors presented? (color, size, position)

Compositional Inventory: Visual Elements

- Media: What visual media are utilized? (photographs, line art, collage, other)
- Informational Value: What is centered? Top /Bottom? Peripheral?
- Visual Composition: What design elements dominate the cover? (lines, shapes, color, borders)
- Framing: How are design elements used to frame the cover?
- Logo: How is the publisher identified? (color, size, position)

Representational Inventory

- Participants /Roles: Who is in the image (race, gender, age)? Provide numbers.
- Pose: How are the participants posed?
- Vectors: What vectors are observed?
- Setting: What setting is included? Abstract or realistic?
- Objects: What objects other than people are included in image?
- Actions: What literacy event (social /literacy practice) is being suggested?

Interpersonal Inventory

- Gaze: Do the characters look at the viewer (demand) or away (offer)? What does this suggest?
- Interpersonal Distance: (close personal, far personal, public)
- Angle of Interaction: Is the viewer positioned from above, below or eye level?
- Modality: Is the image realistic or abstract? How is this created? Is the image posed or naturalistic? (detail, background, focus)

Admittedly, some of the considerations were more relevant than others to our set of book covers, but in general, we found it to be a useful framework, especially when combined with its authors' admonition to always consider these elements in relation to their sociocultural contexts. Serafini et al. support Halliday's recommendation to focus on the text as "a type of social action rather than simply a decontextualized object to be analyzed," pointing out that contemporary multimodal analysis now includes "the social and cultural embeddedness of these texts." They further explain, "No matter the semiotic means of representation, the relationship among visual images, design features, and written language and their associated meaning potentials are socially embedded and worthy of investigation" (98).

Comparative Analysis of Book Covers

Our analysis was also informed by our findings in a previous study (Tsilimeni and Panaou), which was presented in 2014 at The Child and the Book Conference and in which we had identified the main factors influencing Greek publishers' decision-making when it comes to importing international texts. While that study did make use of some quantitative data, its focus was a qualitative analysis of five interviews conducted with three translators and two executives of publishing houses in Greece. The interviewees were Kostia Kontoleon (translator), Ilias Mandilaras (translator), Argiro Pipini (translator), Vaso Papageorgiou (from Metechmio Publications), and Thanos Psichogios (from Psichogios Publications).

All interviewees indicated that nowadays the vast majority of imported children's texts come from English-speaking countries (mainly from the United Kingdom and the United States). The translated covers we analyze here concur with this finding, as they overwhelmingly come from the United States and the United Kingdom. This is indicative but not definitive since our booklist is not comprehensive: The book covers we examine represent a sample of convenience that relatively represents the phenomenon under study (Merriam). Further, while our interviewees had agreed that British and American texts were equally represented in the Greek book market, in our current sample, the American source texts outnumber the British. An additional small number of titles come from Australia, Italy, Spain, Sweden, Finland, and Germany.

The interviewees in our earlier study (see Tsilimeni and Panaou) had attributed the dominance of English-source translations to the following factors:

1. The universality of English
2. Books are imported from places, cultures, and languages that are familiar to Greek publishers
3. Financial issues (the less known a language is the more translation costs there will be)
4. The voluminous production of children's books in the United Kingdom and the United States

5. Cultural relevance (because of globalized mass culture, for example, "Disney-raised" kids are expected to comprehend and relate better to American stories)
6. Aggressive marketing of multinational publishers with increasing influence on the media and collaborations with the film industry.

Dominique Sandis, who is the children's and YA books commissioning editor at Psichogios Publications, was yet another participant from the Greek publishing world that informed our current discussion by reading the shorter (2018 IBBY) version of our paper and providing us with insightful feedback. Commenting on the factors listed above, Sandis emphasized the language barriers, saying that it is only to be expected that publishers would gravitate toward texts they are actually able to read in the original language (usually English).

An experiment we conducted when presenting at the 36th IBBY Congress, is indicative of the globalized mass culture and aggressive marketing factors (numbered as 5 and 6 in the list above). We showed participants the Greek cover for a volume of the *Diary of a Wimpy Kid* and asked the participants who did not speak Greek if they recognized it (Fig. 1).

More than half of the non-Greek-speaking participants recognized the book. *Diary of a Wimpy Kid* is just one example of the multinational corporate world's aggressive global marketing.

Comparing Harry Potter covers brought to a seminar by delegates from Israel, Slovenia, Belgium, the USA, Holland, Germany, Austria, Japan, Hong Kong, Taiwan, Italy, and Spain, Gillian Lathey notes,

> On comparing the covers of these editions and color photocopies of twenty-five international editions taken from the Bloomsbury website, the most striking finding was the predominance of the American as opposed to the original British artwork. Of the twenty-five editions illustrated on the website, fourteen carry the American cover accredited to Mary GrandPré. (144)

Lathey attributes this to the influence and control of Time Warner, which extends far beyond GrandPré's artwork: "Harry Potter, names, characters, and related indicia are copyright and trademark Warner Bros" (144). Lathey explains,

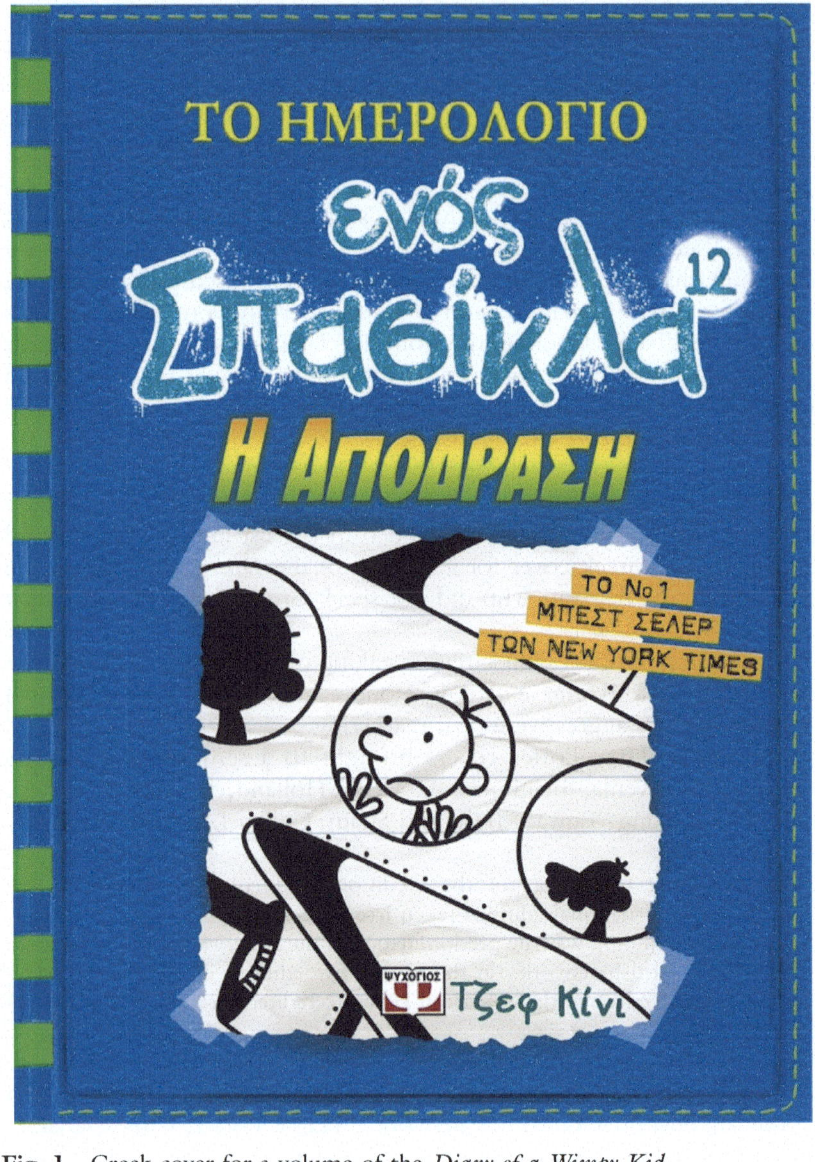

Fig. 1 Greek cover for a volume of the *Diary of a Wimpy Kid*

> Publicity for the films of the series relies, of course, on the visual recognition of GrandPré's designs and the descending lightning-flash at the base of the letter P for Potter. Indeed, this trademark is incorporated into a range of scripts from the Cyrillic (Russian, Bulgarian) to the ideographic (Taiwanese). (144)

She also observes that only the British and Greek editions sport the original cover. Indeed, in some instances, the geographical and cultural proximity of Greece to other European countries might influence Greek publishers' decisions.

Cultural relevance factors are not always related to globalized mass culture, and such factors seem to inform publishers' decisions in choosing to publish a translation. Such seems to be the case with the German-to-Greek translation of *Ich, Zeus, und die Bande vom Olympus* [I, Zeus, and the Olympus Band] (Fig. 2).

Fig. 2 The German original and Greek translation covers of *Ich, Zeus, und die Bande vom Olympus* [I, Zeus, and the Olympus Band]

Not only is Greek children's literature closely attuned with German children's literature—with Greek authors, illustrators, and publishers often participating in book festivals and exhibitions in Frankfurt and Munich, for instance—but the theme of the book is distinctly Greek as well.

As in the case of Harry Potter, geographical and cultural proximity might also influence Greek publishers' decisions when selecting which of the different covers to use, out of the many that the same story may have received in the different languages and countries it was published. For instance, in the case of Rachel Hartman's *Seraphina*, the Greek publisher opted to use the Italian cover. One could claim that what tilted the scales was not so much the proximity with Italy as the fact that they wanted to have a human character visualized on the cover. But then again, they could have used the American paperback version, which does feature a human character (Fig. 3).

Commenting on this particular choice, Sandis, our reader from the Greek publishing world, said that here one should also consider the publisher's subjective aesthetic taste and preference in their choice of cover. The highly original and aesthetically attractive Italian cover, she said, is likely to have been considered more commercial and "catchy" for Greek readers.

Sandis postulated that this cover is comparable to the covers of the highly successful young adult series *Fallen*, and since those books were successfully marketed with a similar cover, then this one could be as well (Fig. 4).

In any case, having images of actual young people on the covers does seem to be a preference for publishers in Greece. When the source cover does not provide that, as in the case of Poblocki's *The Stone Child* (Fig. 5) the publisher might even commission a new cover image altogether.

Sandis asserts that Greek publishers will most often prefer to keep using aesthetically successful covers, or ones that will already be known to the consumer/reader, upon the book's local publication rather than proceed with the origination of a new cover. Therefore, there must have been important reasons for Patakis to commission a completely new cover for *The Stone Child*. Looking at the Greek cover, we infer that, in addition to the preference for illustrations of children, the publisher may also have found the source cover too dark, too scary, or simply too "adult" for a children's book. The latter seems to be the case also for the Greek cover of *La Porta Dels Tres Panys* [The Door with the Three Keyholes], where

Fig. 3 The cover for the American hardcover edition of *Seraphina*, the Greek cover, and the cover for the American paperback edition

Fig. 4 Cover of a volume from the highly successful young adult series *Fallen*

Fig. 5 Source and Greek Translation covers for *The Stone Child*

the Spanish black-and-white image has been replaced with a more colorful and cartoony illustration (Fig. 6).

Is the implied reader of the Greek translation constructed as more stereotypically child-like, then? In our 2014 study, interviewees had identified "age appropriateness" as perceived by the publisher, "compatibility with Greek social and cultural values," and "absence of linguistic and ethical taboos" as important elements that inform decisions about publishing a translation in Greece. All three elements seem to be at play in the changes observed on the translation cover of Anne Cassidy's *No Virgin* (Fig. 7).

Even though the Greek publisher made the risky decision to publish a YA book that resisted the "appropriateness" and "taboo" criteria mentioned above, they tried to tone down these elements by changing the title from "No Virgin" to "I Don't Want To." Keeping a similar cover design, where the second word of the title ("virgin" in the original, "want to" in the Greek translation) is repeated multiple times from top to bottom, the Greek cover alters the message of the source cover.

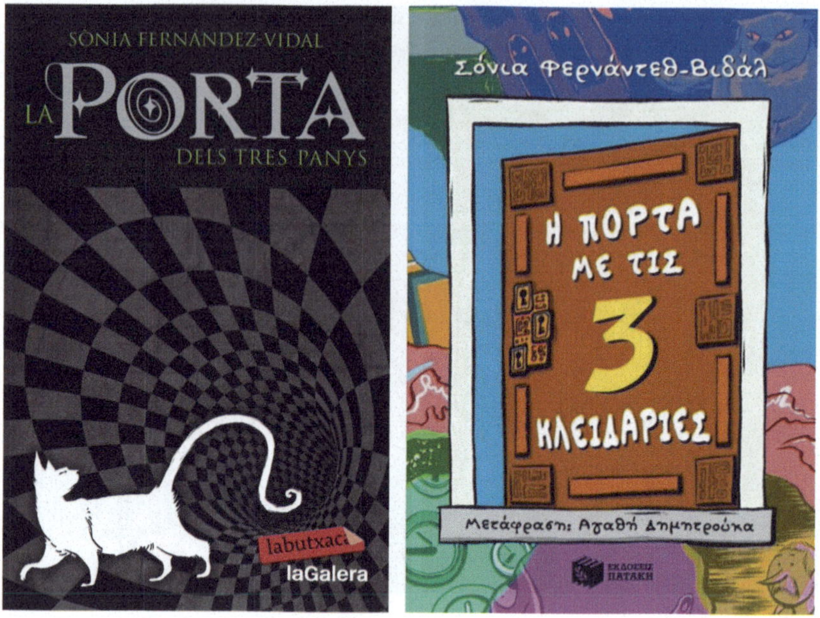

Fig. 6 Spanish original and Greek translation covers of *La Porta Dels Tres Panys* [The Door with the Three Keyholes]

The book now seems to be about a girl torn between wanting and not wanting to have sex, not about a girl who loses her virginity in a violent and traumatic way (after being raped). Sandis, who works at the company that published the Greek translation, explains that it was not the publisher's intention to change or misrepresent the subject of the book. The title was adapted, Sandis says, because its literal translation could be considered "crude" and "unattractive" to Greek educators, critics, and readers. According to Sandis' interpretation, the change from "No Virgin" to "I Don't Want To" could be seen as shedding light on an important subject of the book. According to Sandis, it emphasizes that while the main character wants to have sexual relations with a young man, the moment she says "no" (one of the most significant signals of non-consent in sexual and other situations of harassment), she is not heard and thus raped. Whether a girl says "yes" a hundred times before an act, the moment she says "no" is one hundred times more important, says Sandis.

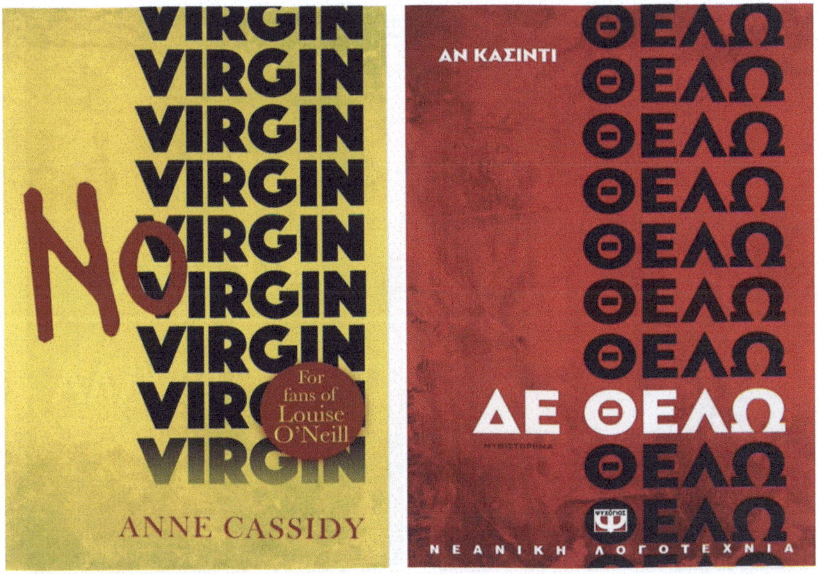

Fig. 7 The original and the Greek translation covers of *No Virgin*

Notably, in the vast majority of the books we have examined, the Greek covers and the source covers look very much alike. Almost fifty percent of the Greek translations (thirty-two out of the sixty-eight) keep the exact same image and layout, with the Greek title being more or less a straightforward translation of the source title. In addition to the Wimpy Kid and Greek Mythology books shown earlier, below are a few more examples (Fig. 8).

In the translation of *A Court of Thorns and Roses*, the title was actually shortened to just "Thorns and Roses" because a word-for-word Greek translation would make the title too long.

The attention payed to fonts and other typographical elements is commendable, as Greek publishers and book designers seem to realize that these typographical elements are important carriers of meaning. Quoting Theo van Leeuwen and his work on typography, Carol O'Sullivan explains that, "[t]he argument could be summarised as 'font also signifies'" (4). She then goes on to argue that, "To date, typeface choice and other printing decisions have not been taken much into account by

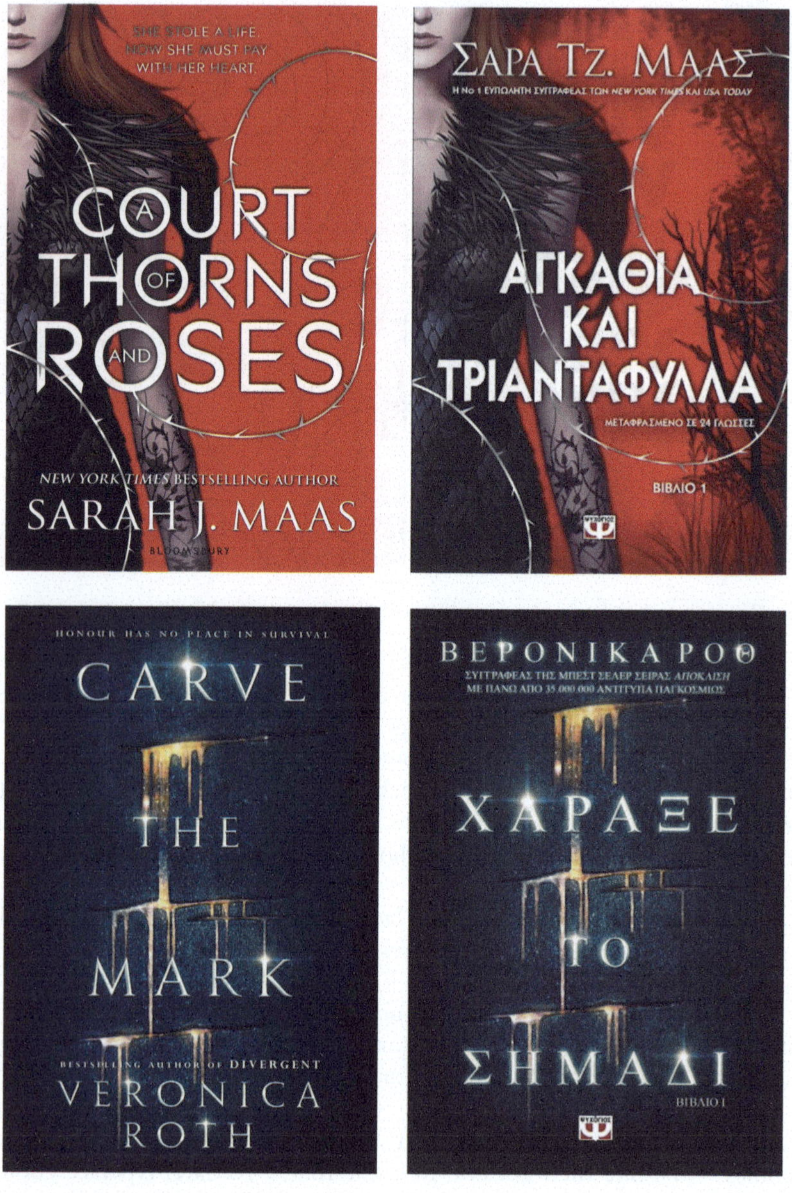

Fig. 8 Source and Greek translation covers for *A Court of Thorns and Roses & Carve the Mark*

scholars of translation. This is perhaps a pity, because typography has, at different times and in different media, been quite active as a translation issue" (4).

We are content when we see such attention to consistency between source and target covers, not because of overt reverence for "the original" but because, as we have argued elsewhere, we value foreignization over domestication. In a chapter in the Routledge *Handbook of Research on Children's and Young Adult Literature*, we use a fusion of narratology and translation studies to analyze Greek translations of picturebooks and make the case for foreignizing constructs of the implied reader of the translation. Márta Minier claims that, "[r]egarding the manner of the translation, the conflict seems to be between making the outcome of the translation process a visibly borrowed text, or rather a familiar sounding one which could have been originally conceived in the receiving language" (102). We argue in favor of "visibly borrowed texts," as young readers should be allowed to experience other cultures than their own through the reading of translated literature. We acknowledge the fact that a translation's implied reader can never—and perhaps *should* never—be identical to the implied reader of the source text; however, we argue for a foreignizing construct of the implied reader of the translation. We favor translated texts whose implied readers have the knowledge, the ability, and the willingness to read and enjoy foreignized texts.

We do recognize, of course, that we cannot judge a translation by its cover; we cannot claim, for example, that books with a foreignizing translation of the cover necessarily feature a foreignizing translation of the rest of the story. Susan Stan has observed such a discrepancy between the covers and content of *Rose Blanche* translations:

> Aside from the French school edition, all of the original hardcover editions of *Rose Blanche* that I have seen (Swiss, German, American, British, Italian, and Spanish) look alike from the outside, with only small variations in the cover finish, yet the views they offer of a little girl's wartime experience are not the same. As the examples given demonstrate, cultural, aesthetic, national, ideological, pedagogical, and economic issues are all at work in shaping these translations. (31)

The Greek translation of *Timmy Failure* is another illustrating example. Both on the cover and in the translated text, the main character, Timmy, keeps his foreign first name, while his last name "Failure" is translated to

"Fiasko," which more or less means the same thing in Greek. However, inside the covers, the rest of the characters are given domesticated, Greek names (Fig. 9).

Sandis states that in Greek translations for children, the localization of names is usually favored in texts addressed to younger ages in order to allow young readers to read "foreign" texts in as familiar a way as possible. Names in such cases are often translated into Greek alternatives, as are other cultural elements that are outside the implied child reader's realm of understanding. However, in the case of translated texts addressed to older children, Sandis continues, foreignization is preferred both by the publishing houses and by the readers, who wish to have the closest possible transfer of the foreign material into their language. This is made possible through foreignizing translations of both texts and covers.

Within our sample of translated covers—admittedly, in their majority addressed to older children—most main characters keep their foreign

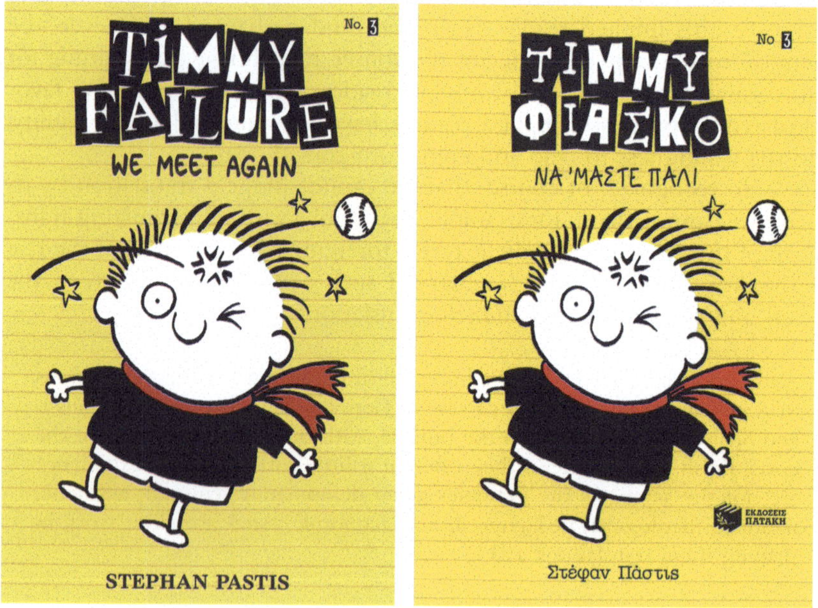

Fig. 9 Source and Greek translation covers for *Timmy Failure*

names either unchanged or with some minor adjustments in the new language (Figs. 10a and 10b).

It seems that the Greek publishers generally did not have much of a choice since they often sign a binding contract with the copyright holders, agreeing to keep the same cover image, layout, and title. In many cases, the Greek translation cover has to even be approved by the foreign publisher before it can go to print.

Even when such legal commitments exist, some subtle differentiations may be observed. The most common ones include the placement and size of the author/illustrator names and the publisher's logo or the addition of marketing elements. Here is an example (Fig. 11) about the placement of a Greek publisher's logo. In the Greek version of *The Boy in the Striped Pyjamas*, the publisher's logo is actually blocking an important part of the cover image. The boys are holding hands, even though they are facing opposite directions; on the Greek cover, the handholding part is blocked by the logo. Thankfully, this was kept only on some digital versions of the cover. Sandis, who works for the publisher of this Greek translation, informed us that they purposefully moved the logo on the printed book, as shown in Fig. 12.

Sandis agrees that the publisher's logo is usually placed in a prominent position on the cover, but she explains that in the case of this particular book, the standard positioning of the logo was intentionally altered in the final printed version so that it would not obscure the cover illustration.

We need to acknowledge at this point that our comparisons have been heavily based on digital images of covers and not on the printed versions of them. This is a shortcoming, not only because the printed versions may differ visually (as in the case of *The Boy in the Striped Pyjamas*) but also because it prevented us from comparing other aspects, such as the size, thickness, glossiness, touch, or quality of the material covers. Stan explains how these aspects can change our reading experience of a translated book: "One can never take for granted that a book in translation will be the same as the original edition, and of course the reading experience can never be duplicated, mediated as it is by the smell of the binding, the weight of the paper, the touch of the cover" (Stan 31).

In any case, our observations regarding the placement of logos are valid. In all of the examined covers, the Greek publishers' logos were highlighted to a much greater extent when compared to the logos on the source covers. This could mean that Greek book buyers pay more attention to, and base their book choices on, who the publisher of the

Fig. 10a Source and Greek translation covers for *Ivy Pocket & The Tapper Twins*

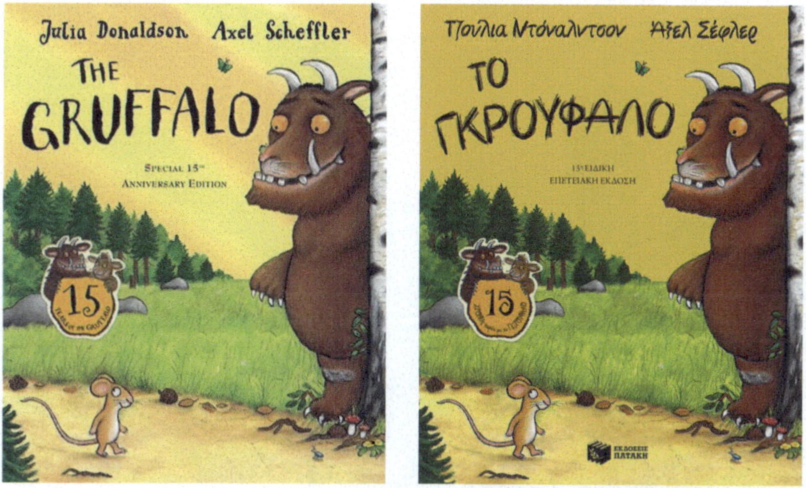

Fig. 10b Source and Greek translation covers for *The Grufallo*

Fig. 11 Source and Greek translation covers for *The Boy in the Striped Pyjamas*

Fig. 12 The printed book cover for the Greek translation of *The Boy in the Striped Pyjamas*

book is. Sandis seems to agree with this inference. This might also explain the Greek publishers' practice to further enhance their branding of some translated covers by indicating that the book is part of a larger-themed or age-group series. The following is such an example, where the added banana icon indicates that the book is included in the "Banana series," addressed to children between the ages of two and four (Fig. 13).

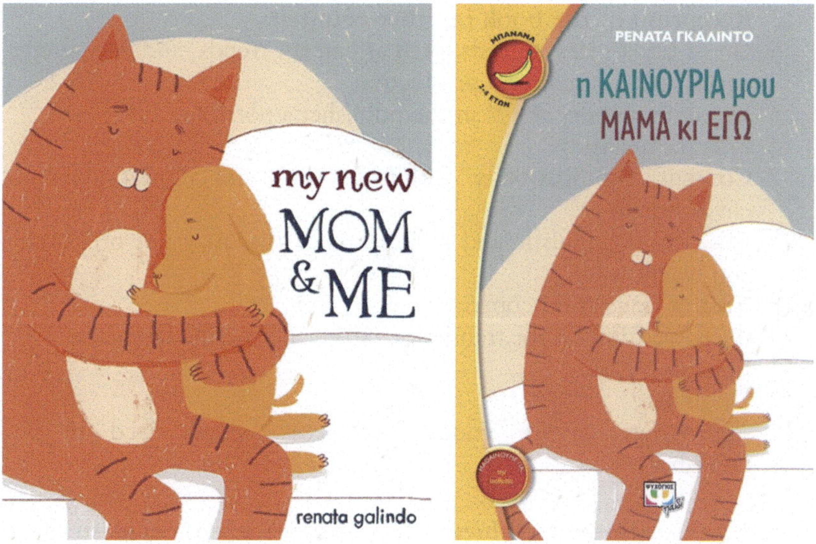

Fig. 13 Source and Greek translation covers for *My New Mom & Me*

Age appropriateness might be another factor that carries more weight in the minds of book buyers in Greece, as might be the books' educational uses. Notice that in this same Greek cover, another "stamp" is added on the bottom left corner, indicating that this story teaches kids about adoption. Sandis purports that this is not a particularly Greek publishing practice and that such series are very popular internationally and feature such intense branding across the board. Labeling information such as the "subject indicator" and so on is added due to feedback from parents, educators, and booksellers in Greece, who wish to have such information handy when looking to buy books from these series.

When it comes to the author/illustrator's name, one can observe the influence of three main selection and marketing factors. The interviewees in our Child and the Book study (see Tsilimeni and Panaou) had identified thirteen different criteria that inform Greek publisher's decision to publish a translation, rating the first three as the most important:

1. **The author's name and recognizability**
2. **Awards received by the author or the book**

3. **Reception of the book in other countries**
4. Greek book market needs, such as current trends, audience preferences, and limited local production
5. Filling in "empty slots" in the publisher's portfolio
6. Literary qualities
7. Content's "compatibility" with Greek social and cultural values and environment
8. Absence of linguistic and ethical taboos that may put off parents and teachers
9. Cost of translation copyrights
10. Available funding for translations from the specific language
11. Engaging plot
12. Originality
13. Appropriateness for young audiences (as perceived by the publisher).

In agreement with the top three criteria, when the author/illustrator is internationally acclaimed, their name is printed in bigger and bolder fonts and placed centrally on the cover. At the same time, big advertisement-like statements highlight the received awards and international success (Fig. 14). The phrase highlighted in yellow at the top of the first cover reads "Now published in 14 countries," while the golden circle/stamp on the second cover reads,

> *New York Times* Bestseller
> Best Book of 2016 according to Publishers Weekly
> Now Published in 17 Countries

For the *Let it Snow* cover, it is indicative that out of the four different covers available in the American market, the Greek publisher opted for the one that highlights the famous author names the most (Fig. 15).

The Greek version also emphasizes the "*New York Times* Bestseller" label by framing it and printing it in red, while the phrase "Now published in 24 countries" is also added inside that same frame. Based on our observations, we could say that Greek publishers seem to think that readers/buyers of translated texts in Greece will select a translation mainly because of its international success and author recognizability.

Fig. 14 Covers of Greek translations that highlight the author/illustrator names, as well as the book's awards and international success

Fig. 15 Four different covers available in the American market for *Let it Snow*

Concluding Remarks

Admittedly, we cannot judge a translation by its cover. But as we have demonstrated in this short presentation, contact and transfer studies can help us examine how cover images, titles, designs, and other elements of the cover are translated/transfered—enabling us to gain useful insights into the transfer and reception of texts for youth in different countries, as well as trends, preferences, and sensitivities in national children's and YA literatures. In his essay in this same volume (pages 117–132), Björn Sundmark demonstrates how the covers for the translated *Lord of the Rings* in different countries and different time periods signal very different audiences, both in terms of age span and in terms of reading expectations.

Bringing our discussion to a conclusion, we would like to join Anna Kérchy and Carol O'Sullivan when they urge us to pay more attention to the role of multimodal elements (such as book covers) in the transfer and reception of cultural artifacts:

> In today's multimodal environment—where the interaction of words with still and moving images, diagrams, typography, page layout, vocality, music, or corporeal performance are "deployed for promotional, political, expressive and informative purposes"—"technical translators, literary translators, copywriters, subtitlers, localisers, publishers, teachers, and other professionals working with language and text" must learn to account for the relentlessly multiplying signifying elements (O'Sullivan, "Introduction" 2) to allow child audiences access to their global heritage, a kaleidoscopic reading experience. (Kerchy 8)

We purport that the book cover of a translated text functions both as the threshold between the real and the imaginary and as the threshold between the local and the foreign; as such, it carries significant weight and meaning(s).

Works Cited

Children's and YA Books

Boyne, John. *The Boy in the Striped Pyjamas*. Random House UK, 2016.
———. *Το αγόρι με τη ριγέ πιτζάμα*. Εκδόσεις Ψυχογιός, 2017.
Cassidy, Anne. *No Virgin*. Hot Key Books, 2016.
———. *Δε θέλω*. Εκδόσεις Ψυχογιός, 2017.
Donaldson, Julia. *The Gruffalo*. Pan Macmillan, 1999.

———. *Το Γκρούφαλο*. Εκδόσεις Πατάκη, 2014.
Fernandez- Vidal, Sonia. *La Porta Dels Tres Panys*. La Galera, 2011.
———. *Η πόρτα με τις τρείς κλειδαριές*. Εκδόσεις Πατάκη, 2013.
Galindo, Renata. *My new Mom & Me*. Schwartz and Wade Books, 2016.
———. *Η καινούρια μου μαμά κι εγώ*. Εκδόσεις Ψυχογιός, 2017.
Green, John, Lauren Myracle, and Maureen Johnson. *Let it snow*. Speak, 2008.
———. *Let it snow*. Speak, 2009.
———. *Let it snow*. Penguin Putnam, 2012.
———. *Let it snow*. Speak, 2014.
———. *Κάνε να χιονίσει*. Εκδόσεις Ψυχογιός, 2017.
Hartman, Rachel. *Seraphina*. Random House Children's Book, 2012.
———. *Σεραφίνα*. Εκδόσεις Πατάκη, 2013.
———. *Seraphina*. Ember Edition, 2018.
Kinney, Jeff. *Diary of a Wimpy Kid—The Getaway (Book 12)*. Penguin Books, 2017.
———. *Το ημερολόγιο ενός Σπασίκλα Νο 12- Η απόδραση*. Εκδόσεις Ψυχογιός, 2017.
Krisp, Caleb. *Anyone but Ivy Pocket!* Bloomsbury Publishing PLC, 2015.
———. *Μακριά από την Άιβυ Πόκετ!* Εκδόσεις Πατάκη, 2016.
Lauren, Kate. *Rapture (Fallen)*. Bantam Doubleday Dell Publishing Group, 2012.
Libenson, Terri. *Η Αόρατη Έμι*. Εκδόσεις Ψυχογιός, 2017.
Maas, J. Sarah. *A Court of Thorns and Roses*. Bloomsbury USA, 2015.
———. *Αγκάθια και Τριαντάφυλλα*. Εκδόσεις Ψυχογιός, 2017.
Pastis, Stephan. *Timmy Failure: Mistakes Were Made*. Walker Books Ltd, 2013.
———. *Τίμμυ Φιάσκο: Τα λάθη είναι ανθρώπινα*. Εκδόσεις Πατάκη, 2014.
Poblocki, Dan. *The Stone Child*. Random House Books for Young Readers, 2009.
———. *Το πέτρινο παιδί*. Εκδόσεις Πατάκη, 2011.
Rodkey, Geoff. *Τα Ταππεράκια κάνουν άνω κάτω τη Νέα Υόρκη*. Εκδόσεις Πατάκη, 2016.
———. *The Tapper Twins Tear Up New York*. Hachette Book Group, 2016.
Roth, Veronica. *Carve the Mark*. HarperCollins, 2017.
———. *Χάραξε το σημάδι*. Εκδόσεις Ψυχογιός, 2017.
Savit, Gabriel. *Η Άννα και ο Χελιδονάνθρωπος*. Εκδόσεις Ψυχογιός, 2017.
Schwieger, Frank. *Εγώ, ο Δίας, και η συμμορία του Ολύμπου*. Ψυχογιός, 2017.
———. *Ich, Zeus, und die Bande vom Olymp*. dtv Junior, 2017.

Secondary Sources

Anagnostopoulos, Vassilis D. "Xeni paidiki logotechnia metafrasmeni stin elliniki glossa kata to telos tou 18ou aiona kai arches tou 19ou aiona." *Diadromes*, vol. 22, 1991, pp. 90–96.

Genette, Gérard. *Paratexts: Thresholds of Interpretation*. Cambridge University Press, 1997.

Kerchy, Anna. "Translation and Transmedia in Children's Literature." *Bookbird*, vol. 56, no. 1, 2018, pp. 4–9.

Lathey, Gillian. "The Travels of Harry: International Marketing and the Translation of J. K. Rowling's Harry Potter Books." *The Lion and the Unicorn*, vol. 29, no. 2, 2005, pp. 141–151.

Merriam, Sharan. *Qualitative Research and Case Study Applications in Education*. Jossey-Bass, 1998.

Ntelopoulos, Kyriakos. *Ta Pedika Anagnosmata ton Pappon mas*. Patakis, 2008.

O'Sullivan, Carol. "Introduction: Multimodality as Challenge and Resource for Translation." *The Journal of Specialised Translation*, vol. 20, 2013, pp. 2–14.

O'Sullivan, Emer. *Comparative Children's Literature*. Taylor and Francis, 2005.

Politis, Dimitrios. "Between Modernism and Postmodernism: Greek Literature for Children and Youth in the Last Decades of the Twentieth Century", *Bookbird*, vol. 56, no. 2, 2018, pp. 52–56.

Serafini, Frank, Dani Kachorsky, and Maria Goff. "Representing Reading: An Analysis of Professional Development Book Covers." *Journal of Language and Literacy Education*, vol. 11, no. 2, 2015, pp. 94–115.

Sheahan, Robyn. "Covers: Windows into Words." *Literature Base*, vol. 7, no. 4, 1996, pp. 26–30.

Stan, Susan. "Rose Blanche in Translation." *Children's Literature in Education*, vol. 35, no. 1, 2004, pp. 21–33.

Tsilimeni, Tasoula and Petros Panaou. "Texts Crossing Time and Space to Reach Children and Youth in Greece: A comparative, Contact and Transfer Study through Interviews to Greek Publishers and Translators." The Child and the Book Conference, 10 April 2014, Athens, Greece. Presentation.

Translating Tenniel: Discovering the Traces of Tenniel's Wonderland in Olga Siemaszko's Vision of *Alice's Adventures in Wonderland*

Karolina Rybicka-Tomala

Translators and translation scholars know that no translation exists in a vacuum. This is especially true when dealing with a well-known work that is part of the children's literature canon and has been retranslated, re-illustrated, reimagined, and reprinted multiple times. Not only can the subsequent translators be influenced by the works of their predecessors, but illustrators of a particular edition may also be influenced—by the original and the translation (be it the one they are illustrating or the one they read as children) or by the pictures in the previous existing editions, sometimes "translating" the original illustrations into the target culture. An interesting example is the case of John Tenniel's illustrations of Lewis Carroll's *Alice's Adventures in Wonderland*, a work often retranslated and re-illustrated. Hendrik van Leeuven (63) and Elżbieta Zarych observe that the drawings accompanying the 1865 edition are still used in numerous editions and retranslation,[1] and their traces can be spotted in the also-plentiful "re-illustrations" of the work. These traces can take the form of composition, ways of portraying the characters (especially Alice), choice of scenes, etc. I will further elaborate on this in a few pages.

K. Rybicka-Tomala (✉)
Jagiellonian University, Kraków, Poland

A. Kérchy and B. Sundmark (eds.), *Translating and Transmediating Children's Literature*, Critical Approaches to Children's Literature,
https://doi.org/10.1007/978-3-030-52527-9_10

In this chapter, I will show the different ways John Tenniel has influenced other artists, using the Polish illustrator Olga Siemaszko as an example. Her case is a particular one, as she has reimagined Carroll's Wonderland not once but at least four times[2]: in 1955; in 1969, when she created two completely different sets of illustrations for the translation by Antoni Marianowicz; in 1964, when she designed a series of six postcards for the centenary of the book; and in 1975, when she created a vinyl cover for a radio-play version of Marianowicz's translation. Each version is distinct in terms of technique, style, and setting; however, each is consistent with the artist's ever-evolving style of the given moment. In each version, we can see that she was inspired by Tenniel, which she herself had acknowledged in an interview for a children's magazine (Wincencjusz-Patyna), yet she had also "polonized" her illustrations (i.e., added Polish cultural elements to them; see Woźniak 302).

First, I shall present Siemaszko's "illustrating strategy" and briefly describe the characteristics of each of her Wonderlands. Then, I shall propose a set of criteria according to which we may compare the "traces" of one artist in the work of another. The last part of the chapter consists of a short case study: a comparison of three scenes using the aforementioned criteria.

Olga Siemaszko and Her Visions of Wonderland

Olga Siemaszko (b. Olga Bielińska in 1911 or 1914, d. 2000) was one of the oldest and most prolific artists of the "Golden Age of Polish Illustration for Children," the period between the end of World War II and the early 1980s (Wincencjusz-Patyna). It was a time when many talented artists turned to children's illustration—mainly because it was an area of most artistic freedom (especially in the Stalinist era) or because (for political reasons) they could not find employment anywhere else and the domain of books for children was deemed "unimportant enough" for them to be able to work.

Along with Jan Marcin Szancer, Siemaszko established *Świerszczyk*, one of the leading Polish magazines for children, in 1945. Until her death in 2000, she had illustrated numerous books, mostly for children. Her style evolved during her long career; every few years, she would develop a variant of her drawing technique and color scheme. She would often return to her favorite works and reimagine them—"translate" them into her new style. *Alice's Adventures in Wonderland* is one such case.

Siemaszko's 1955 version was inspired both by nineteenth-century classics, such as Beatrix Potter's and Kate Greenaway's works, and by Polish folk tradition. The former is represented in the style of drawing the characters, while the latter is evident in Siemaszko placing Wonderland in a meadow full of mostly oversized flowers typical of the Polish countryside and in the Sarmatian dress of the Mad Hatter and Caterpillar. The edition consists of many black and white vignettes and sixteen colored sheets. Further, Alice is dressed similarly to Tenniel's version—in a wide-skirted dress with an apron.

The 1969 version is an example of Siemaszko's inspiration by medieval illuminations and old Persian manuscripts: each image is enclosed in a tightly packed square frame. This Wonderland is full of fine-drawn hills and grotesque figures, all in dimmed, pastel colors. Many illustrations show Alice's point of view, so the heroine rarely appears in the picture. When she does, she is only one small part of the whole scene, a dot in colorful background. Like all characters in this version, Alice's dress resembles the courtly fashions of the eighteenth century, not the nineteenth-century girl's dress of the previous version.

Siemaszko's "extra" versions—namely, the postcards and the vinyl cover (see below)—are quite distinct from the book illustrations. The 1964 postcards are watercolors, presenting a light and breezy Wonderland with a fairy-princess-like Alice in a white dress. The vinyl cover mostly replicates the way the characters are dressed in the 1969 version, yet they are much more detailed and more saturated in color.

Despite the changes, Siemaszko consistently used the strategy of domestication in her illustrations.[3] She used fundamental elements of the Polish iconographic tradition, a trend often observed in Polish illustrations for children of that period (Woźniak 302). For example, in her 1969 Wonderland, all buildings have exaggerated architectural details, such as late-medieval mascarons and baroque domes, reminiscent of the eclectic mix of styles found in her native Kraków, Poland. Her characters (especially the Caterpillar) wear garbs similar to a traditional Orientalized Polish gentry dress, popular since the seventeenth century.

The domestication strategy Siemaszko uses in her illustration is very consistent with the translation by Antoni Marianowicz. However, while the relationship between the illustrations and the translation is an interesting one, this chapter is about the translation of Tenniel's illustrations and not Carroll's text. Therefore, I will just note that Marianowicz's translation is one of the most domesticated versions of all Polish translations, often substituting the poems with well-known Polish nursery rhymes, thus heavily relying on Polish cultural tradition. Michał Borodo notes that "just like Marianowicz's 1955 translation, Siemaszko's illustrations are a domestication of Carroll's novel" (192). I would like to risk another observation: that Siemaszko's illustrations are not only a domestication of *Alice's Adventures in Wonderland* but also—and perhaps foremost—a domestication of Tenniel's interpretation of the work.

From Carroll to Tenniel, and Back Again

Siemaszko's frequent revisiting and reimagining of her version of *Alice's Adventures in Wonderland* almost serendipitously fit into the publishing history of the book. When talking about Tenniel's influence on other artists, it is crucial to point out that his drawings were to a noticeable degree based on the manuscript, *The Adventures of Alice Under Ground*, the digital version of which is now publicly available on the British Library webpage. Carroll's deep involvement in the entire publishing process—including many inspections, instructions, and corrections he made or attempted to make—has been widely described and has been thoroughly described by scholars, such as Michael Hancher and M. P. Hearn, as well as by Zoe Jaques and Eugene Giddens in their work on the publishing history of Alice. Hancher notes that about three-quarters of Tenniel's illustrations were based on Carroll's drawings (28). Hearn, as well as Jaques and Giddens, points out that Lewis Carroll had given "his artist a detailed list of all the subjects to be illustrated, noting also their exact dimensions and placements upon the book's pages," writing to him and frequently calling on his studio (15). Jaques and Giddens notice that "whilst Carroll remained deeply concerned by the production of the illustrations, any changes to Tenniel's artwork tended to be minor. In fact, the book's final images are surprisingly close to the illustrator's initial renderings" (13).

It is also worth noting, that there are many versions of Tenniel's illustrations to speak of. Both Carroll and Tenniel were very particular as to how the images are printed, which led to not one, but three "first editions" of the work, either to be scraped or "sent to the provinces" or discarded for the American market (for which the author did not have much regard). Jaques and Giddens note,

> There are therefore three different 'first' editions of Alice's Adventures in Wonderland: 1) the first UK printing by Oxford University Press with a UK title page, as distributed by Carroll to friends in July 1865; 2) the first UK published edition, printed by Richard Clay and published by Macmillan in November 1865; and 3) the first UK printing by Oxford University Press with an Appleton title page, as published in the United States in May 1866. (19)

Alice's Adventures in Wonderland and its sequel, *Through the Looking-Glass* are not the only authorized Alice books. In 1886, the facsimile edition of the manuscript given to Alice Liddell was published. The same year, a musical pantomime based on both novels premiered on the West End with music by Walter Slaughter and lyrics by Henry Saville-Clark, Lewis Carroll, and Aubrey Hopwood. In 1890, *Nursery Alice*[4] came out—a rewriting by Carroll himself, adapting the source text for younger readers with, as the title page claims, "twenty coloured enlargements from Tenniel's illustrations to Alice's Adventures in Wonderland" that have been not only colored but also slightly altered. Kiera Vaclavik notes that this rewriting was made to fit the prevailing fashions, which have changed since the first editions, at Carroll's insistence (78).

I further explore the relationships between Carroll's illustrations and the evolution of Tenniel's redrawings in my doctorate, yet I do not take them into account in the following case study—as we have no evidence of Siemaszko being acquainted with either the manuscript (or its facsimile edition) or the 1890 adaptation.

Traces of Tenniel

When looking for traces of one artist's inspiration by another illustrator, we can employ a few criteria of similarity. Jan van Coillie uses an analytical model integrating "the central concepts of [Gunther R.] Kress and [Theo] Van Leeuven in the classification by Kris Nauwelaerts for the analysis of picturebooks" (278), distinguishing the following characteristics: illustrative (figuration, setting, etc.), formal (color, contrast, lines, etc.), and graphic (integration of text and illustration, framing, typographical features, etc.). Further, when writing about the elements of the Polish iconographic tradition in illustrating Cinderella, Monika Woźniak distinguishes three "fundamental elements" which can be used more generally in comparing images: "selection, composition, and the contents of the images" (302).

For comparing particular illustrations, I propose a similar distinction. My first criterion is *scene and character selection*, that is, whether the illustrators chose similar scenes to illustrate and whether they portray similar characters in said scene. I have noticed, that although the Wonderland is vast and offers many illustrating possibilities, many editions feature a selection similar to that of Tenniel's in the original edition (see case study below). My second criterion is *composition*, namely, how the characters and objects are placed. My last criterion, *style*, combines Nauwelaerts' formal and illustrative characteristics: it pertains to whether the images are similar in technique and styling.

In broader studies of particular editions, Nauwelaerts' category of the "graphic characteristics" is an immensely useful and significant one. However, I have decided to omit it in this study as I am dealing with illustrations that either do not appear in books or—in the case of the 1969 illustrations—have been reprinted and reused in numerous editions with different overall graphic design.

Case Study: An Analysis of Selected Scenes

In this section, I will present and analyze three scenes that appear in both book editions of Siemaszko's *Alice in Wonderland*, in the series of postcards, and (in the last example) also on the cover of the vinyl. I compare them to Tenniel's illustrations using the abovementioned criteria of similarity. Naturally, they all meet the first one: the selection of a given scene.

The Duchess, Pepper, and the Child/Pig (Chapter 6)

John Tenniel's composition of the scene in which Alice meets the Duchess is such that apart from Alice, who enters the frame from the right, we see four figures: The large-headed Duchess (thought to be inspired by Matsys' The Ugly Duchess) in the center; the crying infant on her lap; the Cheshire Cat, smirking in the low left corner; and the pepper-addicted cook on the left, standing over a cauldron behind the Duchess with her back turned from the rest (Fig. 1).

As I have mentioned before, Siemaszko often leaves out Alice from her illustrations, showing us the scene as if from the heroine's point of

Fig. 1 Tenniel's illustration

view. This is what happens in this scene. However, apart from the absence of Alice, Siemaszko maintains Tenniel's composition: the Duchess in the center (although in her version the child has already transformed into a pig), the cat on left, and the cook peppering her surroundings in the background (Figs. 2, 3, and 4).

The Footmen and the Letter (Chapter 6)

The illustrators of *Alice* often choose to portray a rather short scene from the beginning of chapter six from *Alice in Wonderland*, "Pig and Pepper". The royal Fish Footman delivers a letter to the Duchess, which is received by a Frog Footman. Tenniel deeply anthropomorphizes the features of both servants; apart from their animal-like faces, one might mistake them for humans, clad in elegant livery, and periwigs (the Fish Footman also

Fig. 2 Siemaszko's 1955 illustration

wears a bicorne). In Tenniel's illustration, the Frog Footman stands on the left, on the steps of a deck with a column; both footmen hold a large envelope, and behind them is a forest (Fig. 5).

In all of Siemaszko's versions, the footmen retain more animal-like features. While still in livery and wigs, the Frog has an ovaloid, amphibic shape, and the Fish's tail blends into the livery. We see an unmistakably ichthyic creature with human-like legs, reminiscent of Edward Lear's nonsensical drawings—especially in Siemaszko's most abstract version from 1969. In the first illustration (from 1955), the Fish Footman has fins, instead of arms, emerging from its coat. This version is the most similar to Tenniel's, composition-wise. On the steps on the left stands the Frog Footman, and both footmen hold the oversized envelope. Instead of trees, we see oversized field flowers—a characteristic of Siemaszko's 1955s Wonderland-Meadow (Figs. 6, 7, and 8).

Fig. 3 Siemaszko's 1964 postcard illustration

The Fish Footman also stands in front of the door in both versions from the 1960s (the postcard version also has columns), but in both cases, Siemaszko changes the point of view and the Duchess' residence is in full view. The buildings in both versions are very similar, with chimneys and a dome, in keeping with Siemaszko's "Cracoviasation" of Wonderland. Both 1960s illustrations also show different moments of the scene: On the postcard, the Fish Footman is walking away while the Frog Footman holds the relatively small envelope; on the other hand, the picture from the 1969 edition portrays the moment just before the Queen's servant hands over the invitation.

Advice from a Caterpillar (Chapter 5)

The final scene that I have chosen for the analysis is the heroine's meeting with the hookah-smoking Caterpillar, who is sitting on a mushroom (Fig. 9).

In the Tenniel version, the center of the illustration is filled by the wide-brimmed mushroom. On its cap sits the caterpillar, with its back

Fig. 4 Siemaszko's 1969 illustration

Fig. 5 Tenniel's illustration

turned to the viewer. We see its mysterious, anthropomorphic profile, its predominantly larval features blending with a wide sleeve, and its human-like, gloved hand holding the hookah. The creature blows out the fumes into the upper-right corner. In the lower-left corner, we see

Fig. 6 Siemaszko's 1955 illustration

Fig. 7 Siemaszko's 1964 postcard illustration

two smaller mushrooms within the blades of grass. On the right, Alice stands on tiptoes, barely managing to glance over the cap of the mushroom at the Caterpillar, her wide skirt forming a forth mushroom in the meadow (Fig. 9).

Composition-wise, the 1955 reimagining of the scene is closest to Tenniel. However, the shrunken Alice stands under the giant mushroom. The Caterpillar is posed similarly to Tenniel's, although its head is definitely not anthropomorphized; its shape remains larval. It is clad in striped trousers and a fancy coat and cape, similar to hussar or hajduk livery of the Polish-Lithuanian Commonwealth Army. This is an interesting reinterpretation of the "exotic" and "Oriental" elements present in Tenniel, represented by the hookah and wide sleeves, at once "domesticating" the "Oriental" elements into the "Orientalized" elements of Polish Culture. Instead of the hookah, the 1955 Caterpillar smokes a long-stemmed

Fig. 8 Siemaszko's 1969 illustration

Fig. 9 Tenniel's illustration

churchwarden pipe, which is also an "Orientalized" attribute of the aforementioned cavalrymen. The domestication is also visible in the type of the mushroom—its shape and color pointing to either the saffron milk-cap or chanterelle, both staples of Polish cuisine (Fig. 10).

In the other versions, Alice reaches the cap of the mushroom, also standing on tiptoes on the right side of the image. In the postcard version, the fungus looks like a porcini mushroom, another Polish culinary staple. It still occupies the central part of the picture. The hookah is back, this time placed on the ground in the lower-left corner. The Caterpillar faces the audience. In this version, it still has a cape, though it no longer wears the livery but a coat, vest, bowtie, and nightcap (reminiscent of the shape of the Caterpillar's head in Tenniel's version). It is anthropomorphized (it has a moustache, nose, and arms), but its "legs" swirl as if they were actually a tail.

Fig. 10 Siemaszko's 1955 illustration

It is worth mentioning that this and the following version feature an Alice with a skirt shape similar to that of Tenniel's Alice (Fig. 9).

The scene in the 1969 edition seems to be an amalgamation of the previous ones, translated into the dominating style of this period of Siemaszko's work. The composition is still similar to that of Tenniel's, although the mushroom and figures take up less space in the frame. The Caterpillar resembles the postcard version (Fig. 11), but this time, its legs are not twisted and it is no longer smoking a hookah but a churchwarden pipe (Fig. 12).

Siemaszko also illustrated a cover for a 1975 radio play adaptation by the translator Antoni Marianowicz (Fig. 13). It features Alice and the Caterpillar, but this time, he has a companion—the Mad Hatter. The colors and clothing on the characters resemble those of the 1969 edition, although with traces of a new "hippie-like" style that Siemaszko adopted

Fig. 11 Siemaszko's 1964 postcard illustration

in the 1970s. Instead of toned-down pastels, the tones are more flashy and with deeper contrasts. Of course, this is not an illustration of the scene but a picture of a vinyl cover.

Conclusions

When dealing with subsequent translations, retranslations, reprints, and reeditions of a classic illustrated work, we must remember the relationship of not only the words of the original and translation(s), and the words and pictures, but also of how one set of illustrations may have influenced other artists.

The web of interconnected translations, re-imaginings, and interpretations is especially elaborate and complex in the case of Carroll's *Alice's Adventures in Wonderland*—a classic that has been present in the children's literary canon for over 150 years in numerous translations and interpretations. The case of Olga Siemaszko's multiple visions of Wonderland is a unique one, but in all versions, we can see a consistent fidelity

Fig. 12 Siemaszko's 1969 illustration

Fig. 13 Siemaszko's 1975 vinyl cover illustration

to and inspiration by Tenniel, translated into the Polish iconographic tradition.

Notes

1. Elżbieta Zarych (2020) notes that "for about fifty [Polish] editions of *Alice's Adventures in Wonderland*... at least twenty-three feature the illustrations by Tenniel." (Unless stated otherwise, all translations from Polish are by the author.)

2. In my research, I have stumbled upon an image from a long-ended exhibit of Siemaszko's works that would suggest she had started a fifth series of Alice-inspired images in the late 1980s; however, this is unmentioned in any other works and has never been officially published. As of January 2019, my search for it continues.
3. The domestication of Wonderland seems to be a common strategy of foreign illustrators of *Alice*. For instance, Riitta Oittinen maintains, "While in many of the British illustrated versions, the Gryphon, the Mock Turtle, and Alice are situated in a British landscape at the seaside, in my version I have placed the characters in a more Scandinavian landscape with spruce trees at the background" (131).
4. What is striking about the book, is that it is not an abridgement of the original or merely an intralingual translation, but a translation of an illustrated novel into a multimodal, picture-centric reading experience similar to that of reading a modern picturebook.

Works Cited

Children's Books/Radioplay

Carroll, Lewis. *Alice's Adventures in Wonderland*. Oxford: Oxford University Press, 1865/1998.

———. *Alicja w Krainie Czarów*. Translated by Antoni Marianowicz. Warszawa: Nasza Księgarnia, 1955.

———. *Alicja w Krainie Czarów*. Translated by Antoni Marianowicz. Warszawa: Nasza Księgarnia, 1969/1988.

———. *Alicja w Krainie Czarów* (radioplay). Written by Antoni Marianowicz, music by Ryszard Sielicki, directed by Wiesław Opałek. Warszawa: Polskie Nagrania Muza, 1975.

Secondary Sources

Borodo, Michał. "Przekład, adaptacje i granice wyobraźni." *Przekładaniec*, vol. 28, 2014, pp. 179–194.

Hancher, Michael. *The Tenniel Illustrations to the Alice Books*. Ohio State University Press, 1985.

Hearn, Michael Patrick. "Alice's Other Parent: John Tenniel as Lewis Carroll's Illustrator". *American Book Collector*, vol. 4.3, 1983, pp. 11–20.

Jaques, Zoe and Giddens, Eugene. *Lewis Carroll's Alice in Wonderland and Through the Looking Glass: A Publishing History*. Routledge, 2016.

Oittinen, Riitta. "Where the Wild Things Are. Translating Pictures Books." *Meta*, vol. 48, 2003, pp. 128–141.

Vaclavik, Kiera. *Fashioning Alice: The Career of Lewis Carroll's Icon, 1860–1901*. Bloomsbury Publishing, 2019.

Van Coillie, Jan. "The Illustrator as Fairy Godmother: The Illustrated Cinderella in the Low Countries." *Cinderella Across Cultures: New Directions and Interdisciplinary Perspectives*, edited by Martine Hennard Dutheil de la Rochère, Gillian Lathey, and Monika Woźniak. Wayne State University Press, 2016, pp. 275–295.

Van Leeuven, Hendrik. "The Liason of Visual and Verbal Nonsense." *Explorations in the Field of Nonsense 62*, edited by Wim Tigges. Amsterdam: Rodopi, 1987, pp. 61–95.

Wincencjusz-Patyna, Anita. *Stacja ilustracja: polska ilustracja książkowa 1950–1980; artystyczne kreacje i realizacje*. Wrocław: Akademia Sztuk Pięknych im. Edwarda Gepperta, 2008.

Woźniak, Monika. "Imagining a Polish Cinderella." *Cinderella Across Cultures: New Directions and Interdisciplinary Perspectives*, edited by Martine Hennard Dutheil de la Rochère, Gillian Lathey, and Monika Woźniak. Wayne State University Press, 2016, pp. 296–316.

Zarych, Elżbieta. "Ilustracja w przekładzie. Przekład i ilustracja ilustrowanych książek dla dzieci." *Edytorstwo literatury dla dzieci i młodzieży/Sztuka Edcji*, edited by Marcin Lutomierski, 2020.

Digital Media Transitions

Grammars of New Media: Interactive Trans-Sensory Storytelling and Empathic Reading Praxis in Jessica Anthony's and Rodrigo Corral's *Chopsticks*

Cheryl Cowdy

> "Our job is not to wreck the book but save it
> by teaching grammars of new media."
> Marshall McLuhan[1]

Permitting children to read online "deprive[s] them of an elevating and enlightening experience that will enlarge them as people." So argues Annie Murphy Paul in a 2013 *Time Ideas* article that seeks to defend the "deep reading" of print literature against what she calls the "superficial reading we do on the Web." Readers of books are more apt to "be better able to understand other people, empathize with them and view the world from their perspective," Paul suggests. Citing a 2010 study by Raymond Mar and Keith Oatley, Paul extends this argument about readers explicitly to children: "the more stories they had read to them, the keener the 'theory of mind,' or mental model of other people's intentions." We all know the limits of such discourses of crisis. Since Postman's *The Disappearance of Childhood* (1982), technological change has been

C. Cowdy (✉)
York University, Toronto, ON, Canada

A. Kérchy and B. Sundmark (eds.), *Translating and Transmediating Children's Literature*, Critical Approaches to Children's Literature,
https://doi.org/10.1007/978-3-030-52527-9_11

associated with the end of the book and the ruin of childhood. However, there are two things that intrigue me most about Paul's piece: The first is its emphasis on the disappearance of *empathy* as the new casualty of the dystopian changes threatened by our postmodern engagement with technology. The second is its focus on children as the demographic who might save us; thus, the empathy of young people, like their innocence, is that which is in need of adult protection if children are to continue to represent a better future for us all.

The logic of the first argument contradicts Marshall McLuhan's assessment fifty years ago of the electric/electronic age as a return to the pre-Gutenberg period, when the interplay of our senses meant a more corporeal, empathic engagement with the extensions of human senses and with each other. "In the electronic age," McLuhan suggests, "we encounter new shapes and structures of human interdependence and of expression which are 'oral' in form even when the components of the situation may be non-verbal" (*The Gutenberg Galaxy* 3). Empathy and participation are then interdependent. As "our world shifts from a visual to an auditory orientation in its electric technology," McLuhan predicted a concomitant shift toward "empathy and participation of all the senses" (26, 28). The logical implications of Paul's second position—that young people represent the future of empathy—haunts this chapter and my own attempts to engage young people as co-researchers in a project to explore and compare their reading experiences with a text available both as a book and as an iPad app, Jessica Anthony's and Rodrigo Corral's *Chopsticks* (Razorbill, 2012; Penguin Group USA, 2012).

This paper explores the possibilities of empathic experience created by *Chopsticks*' multisensory, multimedia, multimodal, and transmedia narrative strategies, using as a theoretical framework Marshall McLuhan's theories concerning, "hot" and "cool" media in *Understanding Media*, and the significance of changing "sense ratios" created by the extension of new technologies "into the social world," as he first posited in *The Gutenberg Galaxy*. Exploring the tension between my own textual analysis and the affective responses reported by youth interpreters and by *Goodreads* reviewers of various or undefined ages, I explore how *Chopsticks* invites readers—regardless of their status as adult or young person—to enter "the multimodal subjunctive" (Mackey, "Stepping," *Narrative Pleasures*), to interact with the story within a space of "interpretative possibility" (Zhao and Unsworth 87), compelling consideration of our senses and emotions in the text's interactive meaning-making processes. As Mackey argues, "[t]he subjunctive is not a mode confined to language, although we understand it through words. In its multimodal incarnations, it offers a relatively precise tool for understanding what makes fictions come alive for their interpreters in multisensory ways" (*Narrative Pleasures* 93). Logically, then, the multimodal subjunctive provides a means of understanding the relationship between imagination and empathy. Moreover, an investigation into this process might help counter moral panics based on implicit assumptions about a projected future dystopia in which the disappearance of childhood, the book, and the human capacity for empathy are all falsely connected.

McLuhan: Technology, Corporeality, and Empathy

Looking closely at McLuhan's explorations into the relationship between media and human sensual experience, I am struck by how his work is so very grounded in corporeality. As Eric McLuhan and Frank Zingrone argue in their Introduction to *Essential McLuhan*, "media affect us physically" (8). "Fundamental for McLuhan," argues Janine Marchessault, "is the corporeal experience of how reality is mediated" (120). McLuhan's insistence that the "medium is the message" emphasizes the consequences for human sensual life of electric technology, for it is, he claims, "our most ordinary sense life which creates the vortices and matrices of thought and action" (*The Gutenberg Galaxy* 30). This interplay between the sensual and the empathic offers researchers a more agentic understanding of the relationship between digital storytelling and reader response. As McLuhan explains in *The Gutenberg Galaxy*,

> It is simpler to say that if a new technology extends one or more of our senses outside us into the social world, then new ratios among all of our senses will occur in that particular culture. It is comparable to what happens when a new note is added to a melody. And when the sense ratios alter in any culture then what had appeared lucid before may suddenly be opaque, and what had been vague or opaque will become translucent. (41)

For McLuhan, then, the social or cultural meanings of a given technology—its "message"—while not irrelevant, is completely dependent upon the changing ratios of sensual experience the medium effects within bodies, cultures, and societies: "A theory of cultural change is impossible without knowledge of the changing sense ratios effected by the various externalizations of our senses" (42). McLuhan's work has had a tremendous influence on more contemporary investigations of corporeality and cybersubjectivity that recognize virtuality as an embodied experience. As Jenny Sundén argues, "the virtual does not automatically equal disembodiment" (5). This counters simplistic arguments like Paul's, suggesting instead that we can imagine a more embodied reading praxis in which meaning-making becomes "translucent" across digital contexts and sociocultural divides.

Grammars of New Media: Interactivity and Trans-Sensory Storytelling

The moral panic over the relationships among books, technology, childhood, and empathy seem to be finding currency at the very moment when authors and publishers are experimenting more with multimodal, transmedia texts for young people. Anthony and Corral describe *Chopsticks* as "a multimedia novel which questions the truth of objects, sounds and images" while simultaneously using music and images to "do the work of textual narrative." Anthony and Corral's description of their book as "multimedia" notwithstanding, my analysis emphasizes the multimodal and multisensory properties of the book and the iPad app, while also finding inspiration in Henry Jenkins' term "transmedia storytelling." This term is useful as a way of "talking about convergence as a set of cultural practices" and as a process "where integral elements of a fiction get dispersed systematically across multiple delivery channels for the purpose of creating a unified and coordinated entertainment experience" (Jenkins). Given the term's flexibility and openness to extension, indeed Jenkins concedes "the more we expand the definition, the richer the range of options available to us can be," I invite us to open it further. I want to include in the definition even more emphasis on process by considering "trans-" as a prefix that presupposes the movement between what McLuhan identified as the "hot hyperesthetic" of the "Ear world" and the "cool neutral world" he associated with the domination of the Eye and visuality after the Gutenberg revolution (*The Gutenberg Galaxy* 19).[2] I propose we call this understanding of the extension of a book's emphasis on the visuality of narrative into an app's engagement with touch and sound as "trans-sensory storytelling."

Chopsticks is a text that explicitly, even self-reflexively, experiments with strategies of interactivity and trans-sensory storytelling across two different media—the book and the app—to engage multiple sense experiences in affective meaning-making processes. What appears to be a story of first love is gradually called into question as certain images challenge the veracity of our understanding of the story, along with the emotional and psychological well-being of its teen protagonist, Gloria (aka "Glory"). Both the book and the app make extensive use of photographs to carry the weight of the narrative, introducing a medium often associated with verisimilitude as a form of representation. Indeed, Corral's images play with the cinematic quality of the medium in the introductory pages, which encourage readers to linger on images of trees and fences that subtly establish the text's setting. For instance, in an image of a maple tree in leaf, we can surmise that it is spring (4–5); while the shadow cast by the sun in an image of a fence establishes the time of day and the likelihood (confirmed a few images later) that the story takes place in a city of fenced parks and concrete, such as New York (6–7). In its book form, the narrative evoked by the photographs is deceptively simple. Gloria Fleming is a child prodigy known for her postmodern interpretations of classical music and for blending "early twentieth-century Russian composers with modern rock music," a detail we learn from a newspaper clipping (41). We know that Glory lost her mother at a very young age, a traumatic experience that is only communicated to us by the juxtaposition of Christmas greeting cards from 1999 to 2000, the latter of which depicts a rather devastated looking family of two—Victor Fleming and his daughter—who thank family and friends for their "support during this difficult year" (33). Since that time, Glory leads a rather lonely, regimented life, home-schooled by her father and dedicated to daily piano practice, until Francisco (Frank) Mendoza and his family move into the home next door.

Frank is a talented visual artist who is apparently struggling at a local boys' school, bullied as a Spanish-speaking immigrant from Argentina. Gradually, Glory and Frank fall in love; their courtship is told to us through their exchange of text messages, sketches, photographs, and mixed music CDs, in which Anthony and Corral's musical selections carefully communicate subtle details about the characters and the nuances of their relationship. Regarded as a threat to Glory's career by her father, the love affair grows more intense during an enforced separation while Glory tours Europe. At this point, the prodigy's state of mind seems to become increasingly fragile, until she breaks down in performance, repetitively playing the children's waltz, "Chopsticks." Once she is placed in a rest facility called "Golden Hands," Glory's relationship to reality and Francisco is called into question. Repeated photos of sketches, such as one of the moving van used by Frank's family, substitute Glory's signature for Frank's (50–1, 280); this hint that the drawings are actually Glory's, among many other clues, imply that Frank is in fact a figment of Glory's troubled imagination, a doppelganger upon whom she projects her unacknowledged needs and emotions.

The ambiguities at the heart of the book are largely the result of its reliance on visual images rather than words to tell the story. Readers posting to the *Goodreads* website frequently discuss the possibility that Glory is suffering from a mental illness, and that Frank is in fact a fabrication of her imagination, contradicting the presumed verisimilitude of his photographic image in the text. The majority of the *Goodreads* reviews I have encountered have experienced the narrative only in its book form, and they largely focus on interpreting the visual clues that challenge a simple understanding of the story as a romance. *Goodreads* reviewer "Kay" sums up both the confusion and the intrigue many other reviewers reported in their experiences with the narrative: "Is Frank real? What happens to Glory? Is everything a lie? At the end, I think this book is more about mental illness than it is about teen romance. The pictures tell a half story that make [sic] the reading feel both incomplete & exciting" (12 Feb. 2017).

The mystery is extended through its companion app, inviting an interactive experience of the book's scrapbook narrative style that further complicates the veracity of the images with its touch design and the engagement with sound and moving images. In their social semiotic approach to touch design, Zhao and Unsworth argue that it is "through the physical act of gesturing" that a user "performs an act of meaning-making in the context of the narrative" (95). When readers/users touch some of the photographs on the scrapbook style pages, other photographs and images not included in the book become visible. For example, a double-page spread placed in the book represents pages from the family photo album featuring photographs of Victor's courtship of and marriage to Glory's mother (22–23). Beside a photo of the couple on a motorcycle is an inscription: "'Because our love is wild' –V" (22), suggesting the album pages were assembled by Victor. In the app, readers/users can touch and move the photos. Hidden beneath this photo of the couple is another of Glory's mother alone, hanging a picture of a sea creature (an image that repeats throughout both texts) on a wall of the family home. The discovery of this photo adds a new mystery to my experience of the narrative: who hid the photo beneath the other and why? How does it complicate the representation of the couple's relationship as it is constructed by the placement of photographs made available in the book?

The app also allows users to "shuffle read" the narrative in a random order, functioning much like the tree structure of a *Choose Your Own Adventure* novel, which Marie-Laure Ryan associates with a level two category of interactivity in her classification of user participation in digital narratives. In this level, which Ryan calls "Interactivity Affecting Narrative Discourse and the Presentation of the Story," "relations between lexias [or, 'different paths through which the network could be read as the same story' (42)] can be analogical and lyrical, rather than standing for chronological and causal relations" (44). Although the randomness of the re-ordered pages seems limited to only one new interpretative possibility in my exploration of the app's "shuffle read" option, it certainly changes my experience of the book's story. The photo album pages that depict Glory's parents' courtship, for instance, resonate different affective content and context in the book than when I encounter them in a "shuffle read," where they may be placed differently than in the static form of the book. If the book communicates Glory's emotional dissolution through chronological and causal visual cues, the app's shuffle read option invites a more "analogical and lyrical" reading experience that moves my focus to the affective quality of the protagonist's story.[3]

Indeed, what becomes lucid when we extend our experience with the narrative by engaging with touch, sound, and moving images extended to the app is the *quality* of Glory's affective life. Like panels in Manga, hyperlinks to *YouTube* videos of the songs on CD playlists evoke this teen's mood and her experience of first love (real or imagined matters not). As Rod Munday observes of the function of music in video games, sound contributes much to the environmental experience of multimodal texts, enriching rather than "merely duplicating visual information" (53). By extending the media to include sound and moving images, the text changes the engagement of our sense ratios and hence, of our affective participation in the story. Here, two examples must suffice. The first is of Glory and Franks' first kiss, the affective event that graces the cover of the book and signals its investment in romantic narrative conventions. The addition of moving images to portray such intimate moments heightens our empathic identification with the characters. As Linda Hutcheon explains, modes that "show" move us "from the realm of imagination to direct participation," while "*interacting* is physically and kinesthetically immersive, entailing visceral responses" (22–3). This is also applicable to our participation in the acoustic space of the text, which is extended with the addition of sound and music. When Glory finds a tape recorder in a box that belonged to her mother, the visual representation can only hint at its emotional significance as a material object (206). The affective content of this older mode of sound technology is made more poignant by the addition of a sound file to the photo, in which we hear Glory's deceased mother singing her a lullaby. For Munday, "Music adds meaning to ... stories, either by confirming the visual message, or by resolving the ambiguities in an unclear message" (60). Thus, a new affective "note" is added to the "harmony" of the narrative: the depth of a mother's love and a daughter's unexpressed grief.

Toward an Empathic Reading Praxis

My analysis of *Chopsticks* is complicated by my exploration of adolescent readers' responses to the book and app in ethnographic research. Working with students aged eleven to thirteen in a middle school in Toronto, Canada, in 2014–2015, I employed mixed methods to explore students' responses to both the book and app versions of the text.[4] Initially, I set out to explore how the addition of sound, touch, and moving images in the app changed young people's affective responses to the reading experience of the book. The results of my inquiry were quite inconclusive: seventeen (out of twenty) students said they listen to and watch the music videos and video clips with multimedia texts like *Chopsticks*. With respect to whether or not sound influenced students' emotional connection to the character's story, results were split. Nine felt more emotionally invested, while another nine said it did not affect their emotional connection.

Young people who made use of sound or video clips in their own creative projects were vague about their reasons for doing so: "I chose to use technology because you can add a lot of things to your story, like music or videos, which makes the story more interesting"; "I used sound, and I used it to show examples of why I like the iPad version more than the book because the iPad offered more, like sound." Those who reported a preference for the app gave some of the following reasons for their preference: "I enjoyed using the iPad to read the book since the app uses sound and videos- which makes the book more understandable and interesting"; "I like seeing the videos and listening to what the characters listen to"; and "I enjoyed listening to some of the videos cause [sic] it shows me more about what Gloria likes." Overall, the young people involved in the study had mixed reactions to the differences in their experiences with the book and iPad versions of the narrative. While some enjoyed the novelty and interactivity of the iPad version, most were equally engaged with the book's innovative use of mixed visual media to tell the story.

My research was set on an unexpected path when one of the students, a young woman of twelve I will call "Laura," lost her mother during our project to a sudden heart attack. A piano player herself, Laura's initial interest in the text had to do with her own love of music, particularly classical, and she had been responding enthusiastically to the narrative and Glory's fascination with the children's waltz, "Chopsticks," an important intertext for the novel. After her mother's death, my research assistant and I decided to abandon our own research goals, spending time with Laura in our weekly book club by responding as much as we could to Laura's desire for a quiet space away from her classmates. While she continued to be invested in discussing the text, we were unprepared for the extent to which much of Laura's own inexpressible emotions came to be projected onto her experience with it.

After considering the alternate reading of the novel as a story of mental illness, Laura became almost obsessed with her assessment of Frank and Glory as "ugly" and "stupid," a response to the characters that seemed to contradict my expectation that the interactive quality of the novel might enrich, rather than detract from, a reader's empathic engagement. In one session, Laura asked for permission to doodle in my book copy of the text, which she enjoyed defacing by drawing the words "stupid" and "dumb" on Frank's T-shirt. Expressions of affect can, of course, take disruptive forms, and my team and I came to recognize the very real, visceral responses the story elicited from Laura, perhaps most notably, conflicting emotions such as anger and sadness. In my most memorable session with her, Laura proposed quietly that perhaps Glory was not "crazy" (her word) but "may just miss her mom." Laura's story is a reminder of the ways textual experiences can have consequences for and correspondences with the everyday lives and personal stories of young people, who may project emotions that might make adults uncomfortable, but that may well be functional strategies of expression and resistance to the expectations adults impose on their emotional lives.

Likewise, if we read Glory's "dissolution" less as madness and more as a strategic resistance to emotional dysfunction, then it becomes possible to read the mystery at the heart of the narrative—Glory's disappearance from The Golden Hands Rest Home—metatextually and paratextually as a critique of adultist panic about young people's relationships to digital culture and media. Glory's story is communicated obliquely through visual and auditory images. She makes use of the strategies of trans-sensory storytelling to prevent her own textual and corporeal disappearance and erasure, a "disembodiment" that is threatened not by her relationship to digital technology, but by the denial of her grief and by the very strict, regimented life imposed upon her by her well-meaning father. Both the content of the story and its trans-sensorial narrative strategies operate as a kind of defense against the disappearance of the book and the body through a dual process of sensorial and technological extension.

If, as the visual cues suggest, Gloria is as talented a visual artist as she is a pianist, then we might also attribute her dissociation to the dominance of one mode of sensorial expression over others. It is an uncomfortable state to be in. There are rewards and pleasures, however, to accepting the text's invitation to enter into the multimodal subjunctive, in which "seeming and feeling 'real' involves a subjunctive move into the world of human possibilities and a deitic shift into a point of view other than the interpreter's own" (Mackey, *Narrative Pleasures* 230). In my experience, the ethics of that deictic shift require me as an adult reader and researcher to move towards a more empathic reading praxis with the fictional young person inside the text but, more importantly, with the actual young people outside it.

Acknowledgements The Social Sciences and Humanities Research Council of Canada supported this research. I would also like to acknowledge the contributions of my research assistants, Amelia Ruthven-Nelson and Tina Benigno. My gratitude as well to the teachers and principals and students of H.P.S, especially the latter, who inspired me to follow a much more exciting approach to the study than I had originally imagined.

Notes

1. Quoted in Neill, Sam, p. 311.
2. Fittingly, Jenkins also uses the term "extension" to differentiate transmedia storytelling from adaptation: "Basically, an adaptation takes the same story from one medium and retells it in another. An extension seeks to add something to the existing story as it moves from one medium to another" ("Transmedia 202").

3. *Goodreads* reviewers who experienced the app version of the story enjoyed the potential for interactivity and connectivity, and appreciated that the app facilitates non-sequential reading. For "Michelle," "Chopsticks is like having a conversation with your friend and being able to immediately share a link or video or song in relation to a topic or just randomly" (13 May 2012).
4. In the first year of ethnographic fieldwork in spring 2014, I worked with a small group of students between the ages of eleven to thirteen in a book club that met weekly during the lunch period. The young people experienced both the book and the app with me and a research assistant, sharing their responses informally and responding to the reading event creatively by drawing pictures or creating their own stories. During the second year of ethnographic fieldwork (2015), I was invited by the teacher of a grade eight class to participate in the teaching of a media module in the Language Arts. Students responded to questionnaires about their engagement with the multimedia aspects of *Chopsticks* and completed creative assignments. For the latter, students could choose to create their own multimedia story (of their own creation, or one that filled in the "gaps" of the *Chopsticks* narrative), or to create book reviews comparing their responses to the book and the iPad app. They were asked to use 2 or more media than make use of 2 or more senses (i.e., visual/ sight: drawings, photographs, video, written text; sound: music, voice recording; touch: textured art or interactive element).

Works Cited

Primary Sources

Anthony, Jessica and Rodrigo Corral. *Chopsticks*. Razorbill, 2012.

Penguin Group USA. *Chopsticks Novel. Apple App Store*, Version 1.0, Pearson PLC, 2 Feb. 2012, https://itunes.apple.com/ca/app/chopsticks-novel/id497983366?mt=8. Accessed 27 Oct. 2017.

Secondary Sources

Hutcheon, Linda. *A Theory of Adaptation*. Routledge, 2006.

Jenkins, Henry. "Transmedia 202: Further Reflections." *Henryjenkins.org*, 1 Aug. 2011, http://henryjenkins.org/blog/2011/08/defining_transmedia_further_re.html. Accessed 17 July 2013.

"Kay." Review of *Chopsticks*. *Goodreads*, 12 Feb. 2017, https://www.goodreads.com/review/show/1911305548?book_show_action=false&from_review_page=10. Accessed 26 Oct. 2017.

Mackey, Margaret. *Narrative Pleasures in Young Adult Novels, Films and Video Games*. Palgrave Macmillan, 2011.

———. "Stepping into the Subjunctive World of the Fiction in Game, Film and Novel." *Loading...*, vol. 2, no. 3, 2008, pp. 1–22, http://journals.sfu.ca/loading/index.php/loading/article/view/46. Accessed 02 Oct. 2017.

Marchessault, Janine. *Marshall McLuhan*. Sage, 2005.

McLuhan, Eric and Frank Zingrone. "Introduction." *Essential McLuhan*, edited by Eric McLuhan and Frank Zingrone. House of Anansi, 1995, pp. 1–10.

McLuhan, Marshall. "Understanding Media." *Essential McLuhan*, edited by Eric McLuhan and Frank Zingrone. House of Anansi, 1995, pp. 149–179.

———. *The Gutenberg Galaxy: The Making of Typographic Man*. 1962. U of Toronto P., 1997.

"Michelle." Review of *Chopsticks*. *Goodreads*, 13 May 2012, https://www.goodreads.com/review/show/329033055?book_show_action=false&from_review_page=1. Accessed 26 Oct. 2017.

Munday, Rod. "Music in Video Games." *Music, Sound and Multimedia: From the Live to the Virtual*, edited by Jamie Sexton, Edinburgh UP, 2007, pp. 51–67.

Neill, Sam. "Books and Marshall McLuhan." *The Library Quarterly: Information, Community, Policy*, vol. 41, no. 4 (Oct. 1971), pp. 311–319, www.jstor.org/stable/4306112. Accessed 27 Oct. 2017.

Paul, Annie Murphy. "Reading Literature Makes Us Smarter and Nicer." *Time Ideas*, 03 June 2013, http://ideas.time.com/2013/06/03/why-we-should-read-literature/. Accessed 17 July 2013.

Postman, Neil. *The Disappearance of Childhood*. 1982. Vintage, 1995.

Ryan, Marie-Laure. "The Interactive Onion: Layers of User Participation in Digital Narrative Texts." *New Narratives: Stories and Storytelling in the Digital Age*, edited by Ruth Page, and Bronwen Thomas, UNP - Nebraska Paperback, 2011, pp. 35–62, ProQuest Ebook Central, http://ebookcentral.proquest.com/lib/york/detail.action?docID=1000308. Accessed 27 Oct. 2017.

Sundén, Jenny. *Material Virtualities: Approaching Online Textual Embodiment*, Peter Lang, 2003.

Zhao, Sumin and Len Unsworth. "Touch Design and Narrative Interpretation: A Social Semiotic Approach to Picture Book Apps." *Apps, Technology and Younger Learners: International Evidence for Teaching*, edited by N. Kucirkova and G. Falloon, Routledge, pp. 89–102.

Translated and Transmediated: Online Romanian Translations of Beatrix Potter's Tales

Dana Cocargeanu

Introduction

The impact of digital technology on children's literature has been a topic of interest for children's literature scholarship in recent decades. While some studies have explored the relationship between the new technologies and print children's literature, others have focused on the new digital literary products, such as e-books and literary apps. For example, Eliza Dresang proposes that changes visible in print children's books, such as an absence of linearity and plot closure, are caused by the "interactivity," "connectivity," and "access" that characterize the present "Digital Age" (295). Other authors discuss similar literary features in digital literature, noting, for instance, that in interactive online narratives, stories differ each time they are performed, may have "shared authorship" and do not conform to linear narrative patterns (Pearson and Hunt 274–275; Flewitt 361–365). An interest in the medium-related specificity of digital children's literature has also prompted proposals for suitable analytical tools to explore it. Ghada Al-Yaqout and Maria Nikolajeva revisit picture book theory and analyze digital literary apps for children using aspects such as

D. Cocargeanu (✉)
Independent Researcher, Alumna of Dublin City University, Dublin, Ireland

A. Kérchy and B. Sundmark (eds.), *Translating and Transmediating Children's Literature*, Critical Approaches to Children's Literature,
https://doi.org/10.1007/978-3-030-52527-9_12

performativity, interactivity, and (im)materiality. A materiality approach is also adopted by Ayoe Quist Henkel, who disputes Al-Yaqout and Nikolajeva's claim that apps are immaterial and examines the apps' "physicality," expressed in the interface with which users interact.

Despite this critical interest, the relationship between digital technology and children's literature remains under-researched with regard to online translations of children's books. And yet, online translations are particularly interesting because they exemplify the double mediation of children's books, that is, their transposition into another language (translation), and into another—digital—medium (transmediation). A well-known case is the fan translation of *Harry Potter* in countries where print translations of the latest volumes are not yet available (Ward; Willsher). Usually completed as joint enterprises, through crowdsourcing,[1] the translations are swiftly made available online to wider fan communities.

This chapter contributes to research on the double mediation of children's literature through online translation, by analyzing the online Romanian translations of British author-illustrator, Beatrix Potter (1866–1943). Potter's stories seem particularly suitable for analysis of double mediation, as they have a close connection with both translation and transmediation. Thus, the tales have been extensively translated (into more than fifty languages), Potter's creative use of words and pictures is interesting for translation practitioners and scholars alike, and she participated in the early translation of her stories into French (Cocargeanu et al.). The stories have also been adapted for a variety of media, including animation (*The World of Peter Rabbit and Friends*), apps (Apple's *The Original Tale of Peter Rabbit*), and film (*Peter Rabbit*).

Studies of Potter in translation refer to print translations into languages such as French (Coitit-Godfrey), Japanese (Otsuki; Yoshida), Lithuanian (Urba), Italian (Ippolito), Romanian (Cocargeanu), and Finnish (Ketola). The history of Potter translations is explored by Dana Cocargeanu et al., who find that the earliest published translations are the Japanese (1906) and Italian (1909) ones and comment on the active role played by Frederick Warne, Potter's publisher, in the publication of early translations into several languages (380). Nina Demourova (70) and Kestutis Urba (79)

[1] According to Julie McDonough Dolmaya, crowdsourcing is "Translation performed online by self-selecting internet users, who are often not professional translators but who are collaborating to complete large projects for (virtually) no financial remuneration" (16).

remark on the low number of Potter translations published during the Bolshevik and communist regimes in Russia (1917–1986) and Lithuania (1940–1990) and assign it to the politicized state control of the selection of books translated then.

These and other studies examine the translation of several features of Potter's works. Some authors focus on visual elements, underlining their importance for the tales. Dana Cocargeanu ("Off the Beaten Track," "The Adventures") discusses the essential contribution of the illustrations, format, layout, and typography to the books. She also explains that these elements are altered in translated Romanian editions, due to business-related considerations and different artistic approaches, which affects the narratives and the potential reading experience. Similarly, Nina Demourova explains that economic difficulties after the beginning of the *perestroika* in Russia prevented publishers from reproducing Potter's original illustrations. Instead, they commissioned new illustrations, which resulted, for instance, in a "domesticated" Russian version of Potter's tales (70–71). Other translation challenges connected to the visual elements in Potter's books include choosing whether to verbalize information present only in the illustrations (Ippolito 90–92; Yoshida 78), creating translated texts of a certain length to preserve the close relationship between text and image (Yoshida 78; Coitot-Godfrey 57), and avoiding inconsistencies between translated texts and illustrations (Ippolito 90–92; Cocargeanu, "Off the Beaten Track" 82).

Potter's tales involve many other translation challenges, such as the translation of proper names and cultural references. Studies find that a range of strategies has been adopted by Potter's translators to date, from adaptation to the target culture, to staying close to the originals (Yoshida 78; Coitit-Godfrey 56; Urba 81–82; Otsuki 21, 23; Cocargeanu, "The Adventures" 204–206).

Potter's writing style has also interested scholars such as Ruriko Otsuki, who explains that Potter's concision may tempt translators to use explicitation (21). Indeed, Dana Cocargeanu ("The Adventures") shows that in a 2013 Romanian translation, explicitation is used, and also that Potter's reserved style is made more dramatic. The translation of language register is mentioned by Nina Demourova, who identifies the related problem of translating formal speech through polite pronouns and specific verb forms in Russian; and by Cocargeanu ("The Adventures"), who assigns the differences in register between Potter's tales and the 2013 Romanian edition to the translator's wish to bring the texts closer to a contemporary audience and her indebtedness to the tradition of Romanian children's literature.

Given that Potter participated in the translation of her works, her views on this topic are investigated by Cocargeanu et al., who conclude that Potter was in favor of adapting her tales to their new readership, rather than staying very close to the originals.

As already explained, most of this scholarship deals with print translations of Potter's tales. Although briefly approached by Cocargeanu et al., online Potter translations have not been explored in depth. This is the aim of this paper, which examines the online Romanian translations of Potter's work, posted between January 2010 and December 2014. They include several stories translated by a mysterious blogger, "Franco Sacchetti," on the blog "Beatrix Potter. Traduceri de Franco Sacchetti" [Beatrix Potter: Translations by Franco Sacchetti] (2010)[2]; a translation of *The Tale of Peter Rabbit* by Diana Popescu, posted on her blog, "Povești pentru copii și părinți" [Tales for children and parents] (2013), and one of *The Tailor of Gloucester* by journalist Andreea Dogar, posted on the news website *Great News* (2014). The translations of *Squirrel Nutkin* and *Peter Rabbit*[3] by Sacchetti and Popescu, respectively, are also posted on a website titled *Beatrix Potter*, whose author and posting dates are difficult to identify.

Unlike crowdsourced translations, the Romanian online translations seem to be the work of individuals, rather than of teams. They are also different from the online *Harry Potter* translations, because, as discussed above, Potter's books involve the additional challenge of visual elements. Thus, the online Romanian translations raise significant questions, relevant for the study of contemporary children's literature, which are explored in this study: What factors can prompt or facilitate this type of double mediation enterprise? What is the impact of the online medium on the translations? How do online translations relate to print translations of the same works?

[2] The works translated include *The Tale of Squirrel Nutkin*, *The Tailor of Gloucester*, *The Tale of Mrs. Tiggy-Winkle*, *The Tale of the Pie and the Patty-Pan*, and the less well-known *The Sly Old Cat*, *The Fox and the Stork*, *Three Little Mice* and *The Rabbits' Christmas Party*.

[3] Most titles of Potter's stories begin with "The Tale of…." In this chapter, these words are omitted when referring to story titles, e.g., *Peter Rabbit* is used instead of *The Tale of Peter Rabbit*. *The Tailor of Gloucester* is referred to as *The Tailor*.

Methods

To answer these questions, I combined research into the Romanian context of publication of Potter's tales, particularly the online environment, with translation analysis of the verbal and visual texts. More specifically, when investigating the factors that may have facilitated the posting of the online translations, I researched Romanian internet users' interest in Potter through online searches, using keywords such as "Beatrix Potter," "traducere" [translation], "limba română" [Romanian language], and "povești" [children's stories]. These searches were conducted in 2014.

To analyze the impact of the online medium on the translations and to compare the online and print Romanian editions, I combined analytical tools in children's literature studies, translation studies, and literary studies. The analysis is based on an earlier, extended study which examines the features of Potter's tales that may pose translation challenges, as well as the translation strategies adopted in Romanian translations (1998–2013) (Cocargeanu, *Children's Literature*). I updated some of the findings of that study by analyzing the most recent, online and print, Romanian translations (2014–2017).

The tales examined are those that have both print and online translations, namely, *Peter Rabbit* (four print and one online translation), *Squirrel Nutkin* (two print/one online), *Mrs. Tiggy-Winkle* (one print/one online), *The Pie and the Patty-Pan* (one print/one online), and *The Tailor of Gloucester* (two online/one print).

In *Children's Literature*, I analyzed Potter's books in the context of scholarship on picture books (e.g., Doonan; Nodelman; Nikolajeva and Scott). For the purpose of the analysis, picture books were defined, following Hallberg, as "books with at least one picture on each spread" (qtd. in Nikolajeva and Scott 11). A distinction was therefore made between books in which the significance of the visual component is suggested by its frequent occurrence (picture books) and books which feature it only occasionally and hence grant prominence to the verbal text (illustrated books). In light of this definition, most of Potter's works, including those in the current analysis sample, can be considered picture books. I analyzed the main features of Potter's illustrations, such as layout, framing, media, and colors, and the nature of the illustrations' relationship with the verbal texts. I also explored the significance of other visual elements, for example, format. To examine the image–text interaction, I used the framework proposed by Maria Nikolajeva and Carole Scott. The two scholars identify several types of text–illustration interaction, placed along a continuum between two extremes, "word" ("a text without pictures") and "image" ("a wordless picturebook") (8), including "symmetrical," "complementary," "enhancing," "counterpointing," and "sylleptic." This framework can be applied to Potter's picture books productively, because the typology is related to narrative elements, such as characterization and setting, which are built through the verbal–visual interaction, and which are important in Potter's tales. I concluded that the text and the illustrations interact in mostly complementary ways to give information about the story characters and settings—that is, they contribute different (but noncontradictory) elements which readers must put together, as in a puzzle. In the updated analysis, then, I also considered the main visual features of the tales and their interaction with the verbal texts, in the most recent online and print translations.

In my earlier study, I also examined the translation of several features of the verbal texts, including their read-aloud qualities and stylistic features such as concision, restraint, and a particular use of language register. Among the most important findings was a remarkable similarity in the Romanian translators' approach to language register, namely, the tendency to use more nonstandard Romanian, in contrast with Potter's preference for Standard English. This finding is significant because it challenges accepted wisdom in Children's Literature Translation Studies, namely, that stylistic peculiarities tend to be translated by standard language registers (Lathey 8). Therefore, in the present study I aimed to find whether this approach to the translation of language register is shared by all the Romanian translations, irrespective of the medium through which they were presented.

The Context of Publication of the Online Translations

The posting of the online Romanian translations seems to have been facilitated by a dearth of print translations, a growing interest in Potter and her work, and an increasing availability of internet services in post-communist Romania. These factors are connected to wider political, social, and economic phenomena which occurred in Romania in recent decades, particularly the 1989 fall of the communist regime and European Union membership, and the subsequent increase in contact with Western culture, as explained below.

The publication dates of the Romanian translations of Potter's tales suggest that the online translations were posted to compensate for an almost complete absence of print editions. Whereas in other countries, such as France, Potter translations have continued to be published since the early twentieth century, the earliest Romanian translation dates from 1998 and includes only six stories; the next edition, a collection of four tales licensed by Warne, was published 15 years later, in late 2013. Other print editions followed in 2014,[4] 2015, and 2017. For a relatively long time, then, the only available print translation was the 1998 one. As I explained elsewhere ("Off the Beaten Track"), this edition replaces Potter's original illustrations, features few color plates and small, black-and-white drawings at the bottom of the pages, and changes the original, meaningful relationship between the visual and the verbal texts. The book is seldom mentioned in online comments by Romanian netizens, which indicates a lack of awareness or popularity, possibly caused by the reduced visual appeal. This is illustrated by the comments posted on various blogs and websites (all translations are mine):

> I've looked for her story books but haven't found any in Romanian featuring the original illustrations. (Svety, October 2008)
>
> Finally, there is a book by Beatrix Potter in Romanian! [i.e. the 2013 edition] (Laura Frunză,[5] "Aventurile lui Peter Iepurașul", January 2014)
>
> ... a very old edition [the 1998 one], I came across it by chance... [My daughter] didn't like Beatrix Potter so much – too much text, not enough pictures :(But I've heard from Laura Frunza that a Romanian edition has been published, featuring the original illustrations :) :) :) (Ligia, January 2014)

What the above also suggests is that, despite the dearth of print translations, there was an interest in Potter and a wish for Romanian translations of her works, featuring the original illustrations. Indeed, Romanians were becoming acquainted with Potter and her work in different ways. Whereas

[4] This edition, imported from the Republic of Moldova and translated by Angela Brașoveanu, proved difficult to secure and therefore was not included in the analysis.

[5] Laura Frunză is a translator of children's books.

under communist rule (1946–1989) contact with Western culture was controlled and restricted, the 1989 fall of communism and the 2007 accession to European Union membership enhanced possibilities for such contact and therefore for becoming aware of Potter's originals. This occurred due to the importation of Western products (including books and films), migration, travel, and the internet, as indicated by my online searches of mentions of Beatrix Potter.

Thus, after 1989, Western products, including books, became more common in Romania. For example, in 2011, Potter's original tales were available in the Anthony Frost English Bookshop in Bucharest (established in 2007). In addition, Romania became more accessible to international visitors and Romanians' international mobility increased. In this respect, it is telling that Ioana Bâldea Constantinescu, the initiator and translator of the 2013 edition, first saw a Potter book while travelling in Britain in 2006 (personal communication, July 2014).

A constant rise in internet use in Romania also boosted the chances of becoming aware of Potter's books. The first large-scale access services were launched in 1997 (Barza), and the percentage of Romanians using the internet increased from 3.61 in 2000 to 55.76 in 2015 (International Telecommunications Union). According to my online searches, the internet served as a source of information about Potter, a tool to purchase the English-language tales and a means to further promote her work among Romanians. The earliest online indications of Romanians' awareness of Potter date from September 2006, when a parenting forum member, "Dianami," posted a link to the Project Gutenberg collection of Potter tales. From 2008, positive remarks about the biographical film *Miss Potter* (2007) appeared in Romanian online forums, including statements that some people were searching for further information about Potter and were interested in buying her books, after watching the film (Svety; Frunză, "Cartea de miercuri"; Luckyrock). Others state that they received Potter's books from friends abroad, or bought them online, and some regret the lack of Romanian translations (Vida-Rațiu; Silving). There is also evidence of some awareness of the animated series *The World of Peter Rabbit and Friends* (Pete; Popescu). Romanians' chances of becoming familiar with her life and works were increased by the fact that Potter is mentioned not only in online articles regarding children and children's books, but also in posts about the visual arts (Pete), environmental protection (Sarbu), daily life (Melaroana), and tourism (Gradinaru).

It is against this backdrop of increased interest in and appreciation of Potter and her tales that the actions of the online translators must be understood. Theirs was a more active interest, no doubt motivated by the desire to make the tales available in Romanian, which resulted in the posting, in early 2010 and 2013, of several translated Potter tales. The December 2014 online translation of *The Tailor* is different in that it was posted when the 2013 print edition, which features Potter's original illustrations, was widely available. Nevertheless, that edition did not include *The Tailor* and relatively few of Potter's tales had been translated overall. Although there was a 2014 Moldovan edition of *The Tailor*, this was probably not so widely distributed in Romania. As a consequence, the news website staff could still deem the story to be little-known and a good match for the profile of their series, "Christmas stories which are not so well-known to Romanian children" (my translation).

In conclusion, Potter was not translated into Romanian before the late 1990s, and it is not clear to what extent Romanians knew her works during the decades of communist rule (1946–1989). Nevertheless, in post-communist Romania there were more opportunities to become familiar with Potter and her original books, due to increased contact with Western culture and internet availability. While some internet users posted information about Potter's English-language originals, others made them available in Romanian translation. All these translation enterprises share an apparent wish to introduce Potter to a Romanian audience: they approach her works as little-known to Romanians.

The Impact of Transmediation on the Translated Tales

Presenting the translations through the online medium impacts primarily on their visual elements, which, given the significance of the visual for Potter's picture books, entails a reading experience quite different from that fostered by the originals—even when allowances are made for the mediation entailed by individual reception and linguistic translation.

Firstly, the online translations' paratexts are significantly different from those of the print editions of Potter's tales. Paratexts are usually defined, following Gérard Genette, as the verbal and nonverbal elements which accompany a book's verbal text. What is characteristic for Potter's books, as well as for many other children's picture books (Higonnet 1990; Nikolajeva and Scott 2006), is an interdependence between the main body of text and "paratexts" such as illustrations, formats, and layouts. Even the paratexts of literary apps for children (for instance, icons and opening screens), which may somewhat alter this interdependence (Al-Yaqout and Nikolajeva), remain related to the stories presented in the apps.

However, some of the paratexts of the online Romanian translations serve different purposes and are often unrelated to the tales translated. These paratexts are hyperlinks embedded as texts or pictures. For example, next to each translated story on Sacchetti's blog there are links to the other translations and to the "Notes" (lengthy posts providing details about the original). Popescu's translation of *Peter Rabbit* is surrounded by links to her other posts (children's stories) and social media accounts, as well as advertisements. The *Great News* web page containing the translated *Tailor* features links to constantly updated news articles, with photographs.

As can be seen, through these hyperlinks, additional material is provided, whose range and quantity exceed those allowed by printed books. Thus, if frames in picture book illustrations function as "doorways" onto the fictional world (Nodelman 50), the hyperlinks associated to the online Romanian translations represent windows opening onto a variety of topics, specific to each blog or website. These internet-specific elements can enhance children's online literacy during the process of reading. For instance, children can learn about the interconnectivity of topics or the susceptibility to change of online materials. They might also be encouraged to read other stories for which links are provided. However, the online reading experience can also have negative effects which are less likely to occur when reading books: children's ability to concentrate on a story might be decreased by the paratextual material, and the news headlines and images on the news website may be distressing.

Secondly, the layout of the online translations also fosters a different reading experience from that offered by the original print books. Several of the tales, for instance, *Peter Rabbit*, *Squirrel Nutkin*, and *Mrs. Tiggy-Winkle*, feature Potter's characteristic layout, in which each double spread includes a color illustration on a page, facing a short text on the opposite page. This type of layout, the most "formal and traditional" layout in picture books, creates a specific "visual rhythm, a series of strong beats" (Doonan 85), which can be transmitted to the narrative and the reading experience. Moreover, as Roger Sale argues, it encourages a specific reading pace and a concentration on the text and image on each double spread (127–128). The shortness of the text, the large amount of white space surrounding it and the fact that each short text has a corresponding illustration emphasize the importance of both, therefore encouraging readers to pause and consider them carefully.

By contrast, the translations published online are presented as vertical sequences of units of text and illustrations, unfolding when readers scroll down the page. As a result, the reading experience is more linear and the narrative does not have the rhythmic regularity of the originals. In fact, the absence of page breaks in the web page format similarly affects even the stories featuring different layouts, such as *The Pie and Patty-Pan*. Whereas page breaks in Potter's picture books can increase suspense (as in *The Pie and Patty-Pan*, 33–41) or make the narrative progress at an appropriate pace (Bartow Jacobs 360), the linear format of the online versions fosters a potentially faster reading experience, with diminished suspense.

Thus, the online Romanian translations remain unapologetically digital in a rather basic way. Their visual design does not attempt to imitate the experience of book reading, for example, by using animation to create the illusion of page-turning; nor is there any indication of an attempt to provide opportunities for readers to interact with the stories in ways similar to literary apps for children. There is little, if any, interactivity and performativity, which, according to Al-Yaqout and Nikolajeva, characterize digital literary products for children. This feature can, most probably, be assigned to differences in resources, skills, and objectives between the Romanian blogs/websites and corporate producers of digital literature for children.

Nevertheless, the size, number, and placement of the illustrations are indeed edited. Sacchetti places the original units of text either above, or above and below the corresponding illustrations, which indicates a concern to preserve the close original relationship between text and picture. A similar concern appears in Popescu's translation, in which the pictures are placed to the right or left of units of text. However, both Popescu's and Dogar's translations dispense with some of the illustrations, which further increases the pace of reading. In addition, Popescu sometimes edits the illustrations to leave out the outer elements and generally to focus on the characters. The accumulated effect of these changes is that the illustrations' contribution to building the setting of the story is reduced. There is less visual information about the forest where Peter lives and Mr. McGregor's garden, because several pictures where they are shown are deleted, and because they are edited out of other illustrations.

As a result, the setting becomes less recognizably Potteresque and more emphasis is placed on the events of the story and the characters.

Furthermore, the illustrations' complementary role in characterizing the protagonists is similarly reduced. For example, as shown by Mullins (2–3), Peter's rebellious nature is suggested by the picture opposite the opening of the story, where all the rabbits' faces are visible, except for one baby rabbit which is showing only the hind part of its body. Similarly, in the picture accompanying Mrs. Rabbit's warning not to go into Mr. Mc Gregor's garden, the characters' grouping and positioning in relation to each other, the direction of the planes in the illustration also indicate Peter's different character: his female sisters are together, facing their mother, whereas Peter is pictured separately, his back turned to them and seemingly moving rightward, impatiently. In the Romanian online translation, the former illustration is edited and the rabbit which could be Peter is not shown, and the latter illustration is removed, which delays the moment when readers are given information about Peter's character.

In addition, the symbolism of clothes in the story is also diluted. Clothes are part of an animal–human duality that characterizes Potter's animal protagonists, accomplished through combining accurate animal features and realistic human traits (Nikolajeva and Scott 94). In Potter's tales, clothes and other human behavior, such as walking on hind legs, are often a symbol of the civilizing—and restricting—adult world. This view of the relationship between children and adults has been linked to Potter's restrictive family environment, Victorian child-rearing practices, and the more general social emphasis on self-restraint and civilized behavior, visible also in the rigidity of clothing (Scott; Bruscini). Nevertheless, it has also been pointed out that clothes can be interpreted as a symbol of growing individuality and identity-building, for example, in Peter Rabbit's experiences throughout *Peter Rabbit* and *Benjamin Bunny* (Mullins). In Popescu's translation, this symbolism is altered by the deletion of several illustrations which originally showed Peter's transformation from a "civilized" rabbit walking on his hind legs and wearing human clothes and footwear, to a real wild rabbit which runs "faster" when he gets down on all fours and sheds his shoes and jacket. This deletion is also supported on the verbal level in the Romanian translation, where the increase in speed is not mentioned: the sentences "he ran on four legs and went faster, so that I think he might have got away altogether" are translated by "a fugit in patru labute, cat de repede putea pentru a scapa de urmaritor" [he ran on four little paws, as fast as he could to get away from the pursuer].

In addition to the visual changes entailed by the transmediation of the translations, the Romanian online translations also raise issues related to authorship and the stability/mutability of online material. As mentioned earlier, Sacchetti's and Popescu's material are used on a website dedicated to Beatrix Potter, with no acknowledgement. Extracts from Popescu's translation are also included in other posts related to Potter or children's books, such as those on the blog "Last Night's Music" and the website *Parinti.com*. Regarding the ephemerality of online material, Ghada Al-Yaqout and Maria Nikolajeva have remarked that literary apps for children can undergo updating, as a result of which initial versions become inaccessible. Somewhat similarly, Sacchetti's blog has disappeared from the blogosphere in recent months, and no trace has remained of the translations, except the unacknowledged one on the website *Beatrix Potter*, unless individuals have saved them for themselves.

In conclusion, the online environment and digital technology have allowed the Romanian translators liberties which authors of print translations can seldom afford. The layout and illustrations were altered based on their tastes, digital skills, and objectives, with consequences for the reading experience, the illustrations' contribution to the narrative, and thus the narrative itself. Furthermore, the online format provided opportunities for them to include information and material related to their work, interests or business, which would not be available to individual translators of print editions. The quantity and range of this material also surpass those that publishers of print editions can include on book paratexts. All this exemplifies how the internet can impact on the translation of children's books, bringing to readers literary products which are not only culturally and linguistically mediated, but also bear the mark of their new medium.

Similarities Between the Online and the Print Translations

And yet, the relationship between the print and online translations of the same works is more complex, as they can also be similar in other respects. Despite the medium-related specificity of the print and digital Romanian translations, they are alike in that they all modify the original visual elements and share translation strategies.

Firstly, it is not only the online translations that alter the visual features of Potter's tales. The Romanian print editions also modify the original small format and text-facing-picture layout, and the illustrations. Thus, in all editions, individual pages feature two or more of the original pages. As I explained elsewhere ("The Adventures," "Off the Beaten Track"), such changes modify the pace of the reading experience and the narrative rhythm, to some extent similar to the online translations. Furthermore, the 1998 edition replaces the original illustrations with new ones, created according to a different artistic approach; the 2015 edition reproduces Potter's pictures in black-and-white; and the 2017 translation is a replica of a pirated 1916 American edition of *Peter Rabbit*, illustrated by Virginia Albert.

All this can alter the original text–image relationships, as well as the characters' features and the setting of the narratives. For instance, in the 1998 and 2017 editions, the animal characters' human–animal duality is diluted. In the former, the illustration which shows Peter running away after shedding his coat emphasizes Peter's human traits, unlike the original, which focused on his animal features. Peter's face is drawn in a Bugs Bunny fashion, he is smiling widely and, although his coat seems to have flown away, he retains undergarments such as shorts and a t-shirt and runs on his hind legs. Thus, the illustration does not enforce the human–animal tension as in the original and even contradicts the verbal text, which stays close to Potter's representation of Peter's feelings of discouragement and fear while being chased by Mr. McGregor. In the 2017 edition, Virginia Albert's animal characters are not as anatomically accurate as Potter's; in addition, a jacket-less Peter Rabbit is shown on different pages, even before the moment when he actually sheds his coat in the narrative. This decisive moment, which can symbolize Peter's return to a more natural, less civilized self (Scott) is therefore diluted on the visual level in the 1916 American and the 2017 (and 1998) Romanian editions.

The above suggests that, unless Romanian children have access to Potter's English-language originals, to date they have known her tales only through such visually modified editions. Therefore, the verbal-level mediation is supplemented by visual mediation in both the print and online Romanian translations, which will no doubt impact on the reception of Potter's works in Romania.

Secondly, the Romanian translators approach Potter's works in similar ways with respect to the translation of the language register. Potter's writing style involves particular uses of language register, which can be challenging in translation due to mere language differences (for instance, different ways to mark formality), or to different ideas regarding the appropriate style for children's books. Thus, in the books analyzed, most of Potter's language is Standard English, and informal language is not used very often. There are occurrences of dated language, such as vocabulary ("most dreadfully frightened," *Peter Rabbit* 17) and hyphenation ("key-hole," *Squirrel Nutkin* 24), as well as formal language and less common phrases, such as "implored him to exert himself" (*Peter Rabbit* 36) and "excessively impertinent" (*Squirrel Nutkin* 16). The high register language reflects Potter's social background, as well as her intention to satirize some of its flaws; but it has also been said to have been inspired by the earlier period of the Regency, rather than late Victorian times, and related to a parodic use of moral tale language (Carpenter). Finally, Potter also uses archaic verb forms ("hath," "sufficeth") and interjections ("alack") in *The Tailor*, to recreate the atmosphere of Regency England, where the story is set.

Language register is used in characters' speech to suggest social status and relationships. For example, formality is increased in the squirrels' speech in *Squirrel Nutkin*: "Old Mr. Brown, will you favour us with permission to gather nuts upon your island?" (15); in *The Pie and the Patty-Pan*, the cat Ribby tends to use pretentious language when talking to herself ("a most genteel and elegant little dog; infinitely superior company to Cousin Tabitha Twitchit"), and Standard or formal English when addressing her tea party guest, the dog Duchess ("I disapprove of tin articles in puddings and pies. It is most undesirable"). In contrast, Duchess tends to use more Standard English and more contracted verb forms when talking to herself ("Why shouldn't I rush along and put my pie into Ribby's oven when Ribby isn't there?"), which indicates that she is less "genteel" than the cat—a conclusion the latter also reaches by the end of the story.

The shared feature of the Romanian translations is a tendency to use nonstandard registers, even when Standard or formal English are used in the originals. The registers of Romanian are categorized into "cultivated language" (standard Romanian, specialized languages and the language of the arts, including that of literature), and "popular" language (i.e., "of the people," "not cultivated") (Zafiu). According to Zafiu, there are a rural and an urban "popular language," the latter comprising familiar (colloquial) language and slang. The Romanian translations use a combination of registers including, to various degrees, "popular language" (colloquial and rural), alongside "cultivated" Romanian (standard and literary).[6] For example, most Romanian translations refer to Peter Rabbit's hiding in the water can by the verb "a se piti," which is a "popular" Romanian version of "to hide," and "very damp" ("[Peter] was very damp with sitting in that can") is translated by "ud leoarcă," "popular" for "soaked wet." The narrator's and characters' speech in the Romanian translations of *The Pie and the Patty-Pan* also includes more "popular" language. For instance, "Duchess ran home feeling uncommonly silly" (54) is translated as "Ducesa a luat-o la fugă spre casă ruşinată că făcuse figura prostului" by Sacchetti, and "Ducesa o zbughi spre casă, simţindu-se ca şi cea mai mare nătăfleaţă" by Florescu (114). In these sentences, the expressions used for "run" ("a o lua la fugă", "a o zbughi") and "uncommonly silly" ("a face figura prostului," "nătăfleaţă") are colloquial.

The Romanian translators' use of "popular language" indicates a shared understanding of the desirable style of children's books. "Popular" Romanian results in more colorful, lively texts, in contrast with Potter's simpler vocabulary and reserved style. The shared conception apparent here is that children's stories should stay close to their readership through the use of colloquial and standard language, and a lively style. This is an indication that the stylistic norms governing the translation of children's literature in Romania may be different from norms in countries where stylistic peculiarities are translated by more standard language registers, discussed, for instance, by Vanessa Joosen (70) and Isabel Pascua-Febles (120). It may be, therefore, a form of domestication which, interestingly, Potter herself favored in relation to the French translation of her tales (see Cocargeanu et al. 381). Furthermore, rural "popular" vocabulary can be perceived as acceptable for children's books, as it features in the nineteenth- and early twentieth-century Romanian children's classics and at present might carry the prestige of high literature (Cocargeanu, "The Adventures" 209–210).

On the narrative level, when popular Romanian is mixed with standard and literary Romanian, the narrator's voice becomes similar to that of narrators of Romanian children's classics, while also sounding like the voice of an educated adult who is not afraid to use colloquial language

[6] This tendency is not very strong in Dogar's 2014 translation of *The Tailor*, which uses mostly standard Romanian; however, there are also some instances of popular language in this translation.

when telling stories to children. As for the characters' features and relationships in the Romanian translations, these should be discussed on a case-to-case basis, as they are not uniformly dealt with. For example, the squirrels' deference to the owl in *Squirrel Nutkin* is rendered through Romanian formality markers such as polite pronouns and plural verb forms which achieve a similar effect to the original text; on the other hand, the addition of "popular" language to the way Duchess and Ribby address each other in the Romanian *Pie and the Patty-Pan* indicates a slightly more relaxed, or less distant relationship than in the English version.

To conclude, the similarities between the online and the print Romanian translations of Potter's tales suggest that some translation approaches are not directly influenced by the medium used (print or online), but may rather be the result of other factors. Certainly, the changes brought to the original visual elements in the online Romanian translations stem from the specifics of digital technology. However, the print editions do not offer perfect visual replicas of Potter's books either, for reasons such as commercial considerations on the part of publishers. Furthermore, the similarities in the translation of language register indicate shared ideas among Romanians regarding the desirable language of children's literature, despite the fact that the translations were produced in different contexts, by bloggers, publishers and even a news website.

Conclusion

What do the online Romanian translations tell us about the relationship between digital technology and children's literature, especially in translated form?

First of all, the internet can contribute significantly to popularizing an author in a country with no tradition of translating that author's works. In Romania, Beatrix Potter was not translated before the late 1990s, and the first edition that featured her original illustrations was published in late 2013. However, Romanian internet users had been posting material and comments regarding Potter several years before 2013, and some of them took on the challenge of making her tales available online in Romanian translation. Furthermore, an increasing interest in Potter's work was also facilitated by other favorable circumstances, namely, the cessation of the limitations on contact with the Western world, imposed by the communist regime. International travel, migration and increased access to Western cultural products, including Potter's original books and the biographical film *Miss Potter*, contributed to raising awareness of her works.

Second, presenting the Romanian versions of Potter's tales through the online medium adds a new layer of mediation to the translation enterprise, impacting primarily on the translations' visual elements. The layout, and the size, number, and placement of the illustrations are altered; hyperlinks are added, which are often not strictly related to the tales translated. All this makes for quite a different reading experience from that encouraged by the originals, and sometimes reduces the illustrations' contribution to the narrative.

Finally, the shared features of the Romanian translations suggest how digital and print translations of children's books can be connected. Both the online and print versions modify the original visual elements of Potter's books, which is bound to affect the reception of the tales in Romania, because to date Romanian children have not been exposed to translations that accurately reproduce the visual–verbal relationships in Potter's originals. In addition, similar strategies are also visible in the verbal text of the translations, notably the tendency to use "popular" (colloquial and rural) Romanian more frequently than Potter used informal English. This strategy brings the texts, especially the narrator's voice, closer to their audience by echoing everyday language and the language of Romanian children's classics, and occasionally decreases the formality in the characters' speech and relationships.

Works Cited

Beatrix Potter Tales

Potter, Beatrix. *The Tale of Peter Rabbit*. The Saalfield Publishing Co., 1916.
———. *The Tale of the Pie and the Patty-Pan*. 1905. Frederick Warne, 2002.
———. *The Tale of Mrs. Tiggy-Winkle*. 1905. Frederick Warne, 2002.
———. *The Tale of Squirrel Nutkin*. 1903. Frederick Warne, 2011.
———. *The Tailor of Gloucester*. 1903. Frederick Warne, 2002.
———. *The Tale of Peter Rabbit*. 1902. Frederick Warne, 2011.

Romanian Translations of Beatrix Potter's Tales

Bâldea Constantinescu, Ioana, translator. *Aventurile lui Peter Iepurașul* [The Adventures of Peter the Little Rabbit]. By Beatrix Potter, Editura Arthur, 2013.

Braşoveanu, Angela, translator. *Povestea iepuraşului Peter* [The Adventures of the Little Rabbit Named Peter]. By Beatrix Potter, Editura Cartier, 2014.

———. *Croitorul din Gloucester* [The Tailor of Gloucester]. By Beatrix Potter, Editura Cartier, 2014.

———. *Povestea veveriţei Nutkin* [The Tale of the Squirrel Named Nutkin]. By Beatrix Potter, Editura Cartier, 2014.

Dogar, Andreea. "Croitorul din Gloucester" [The Tailor of Gloucester]. By Beatrix Potter, *Great News*, 17 December 2014, www.greatnews.ro/povesti-de-craciun-croitorul-din-gloucester-de-beatrix-potter. Accessed 21 March 2017.

Florescu, Andreea, translator. *Aventurile lui Peter Iepuraşul* [The Adventures of Peter the Little Rabbit]. By Beatrix Potter, Editura Andreas, 2015.

Pogonici, Mihaela, translator. *Povestea lui Peter Iepurilă* [The Tale of Peter Rabbit]. By Beatrix Potter, Editura Paralela 45, 2017.

Popa, Ecaterina and Oana Popa, translators. *Aventurile iepuraşului Peter* [The Adventures of the Little Rabbit Named Peter]. By Beatrix Potter, Editura Dacia, 1998.

Popescu, Diana, translator. "Povestea iepurasului Peter" [The Adventures of the Little Rabbit Named Peter]. By Beatrix Potter, *Povesti pentru copii*, 4 February 2013, www.copiipovesti.blogspot.ie/2013/02/iepuras-peter-beatrix-potter.html. Accessed 19 May 2014. Also on *Beatrix Potter*, n.d., www.beatrixpotterandme.weebly.com. Accessed 20 April 2017.

Sacchetti, Franco, translator. "Povestea Veveriţei Ronţăie-Alune" [The Tale of Squirrel Nibble-Nuts]. By Beatrix Potter, *Beatrix Potter. Traduceri de Franco Sacchetti*, 8 February 2010, www.povesti-cu-animale.blogspot.ie/2010/02/povestea-veveritei-rontaie-alune.html. Accessed 19 May 2014. Also on *Beatrix Potter*, n.d., www.beatrixpotterandme.weebly.com. Accessed 12 February 2018.

———. "Croitorul din Gloucester" [The Tailor of Gloucester]. By Beatrix Potter, *Beatrix Potter. Traduceri de Franco Sacchetti*, 30 January 2010, www.povesti-cu-animale.blogspot.ie/2010/01/croitorul-din-gloucester.html. Accessed 19 May 2014.

———. "Vulpea şi barza" [The Fox and the Stork]. By Beatrix Potter, *Beatrix Potter. Traduceri de Franco Sacchetti*, 29 January 2010, www.povesti-cu-animale.blogspot.ie/2010/01/vulpea-si-barza.html. Accessed 19 May 2014.

———. "Povestea Doamnei Strînge-Lucruri-Pierdute" [The Tale of Mrs. Collects-Lost-Things]. By Beatrix Potter, *Beatrix Potter. Traduceri de Franco Sacchetti*, 29 January 2010, povesti-cu-animale.blogspot.ie/2010/01/povestea-domnei-stringe-lucruri.html. Accessed 19 May 2014.

———. "Trei şoricei" [Three Little Mice]. By Beatrix Potter, *Beatrix Potter. Traduceri de Franco Sacchetti*, 28 January 2010, www.povesti-cu-animale.

blogspot.ie/2010/01/trei-soricei-1890-trei-soricei-mititei.html. Accessed 19 May 2014.
———. "Cum îşi petrec iepuraşii Crăciunul" [How the Little Rabbits Spend Their Christmas]. By Beatrix Potter, *Beatrix Potter. Traduceri de Franco Sacchetti*, 28 January 2010, www.povesti-cu-animale.blogspot.ie/2010/01/cum-sarbatoresc-iepurasii-craciunul.html. Accessed 19 May 2014.
———. "Pisica bătrînă şi vicleană" [The Old and Sly Cat]. By Beatrix Potter, *Beatrix Potter. Traduceri de Franco Sacchetti*, 28 January 2010, www.povesti-cu-animale.blogspot.ie/2010/01/batrina-pisica-sireata-1906-aceasta.html. Accessed 19 May 2014.
———. "Povestea unei budinci şi a unei forme" [The Tale of a Pie and a Mould]. By Beatrix Potter, *Beatrix Potter. Traduceri de Franco Sacchetti*, 27 January 2010, povesti-cu-animale.blogspot.ie/2010/01/blog-post.html. Accessed 19 May 2014.

Online Romanian Articles About Beatrix Potter

Dianami. "Povestile lui Beatrix Potter." *Despre copii*, 24 September 2006, www.forum.desprecopii.com/forum/topic.asp?TOPIC_ID=84744&. Accessed 13 May 2014.
"Flopsy, Mopsy si Peter iepurasul." *Last Night's Music*, 1 October 2014, www.lastnightmusic.wordpress.com/2014/10/01/flopsy-mopsy-si-peter-iepurasul/. Accessed 16 August 2017.
Frunză, Laura. "Cartea de miercuri – Povestea lui Peter Rabbit." *Idei pentru mămici si copii*, 7 April 2010. www.laurafrunza.com/2010/04/07/cartea-de-miercuri-povestea-lui-peter-rabbit/. Accessed 13 May 2014.
———. "Aventurile lui Peter Iepuraşul." *Idei pentru mămici si copii*, 22 January 2014, www.laurafrunza.com/2014/01/22/aventurile-lui-peter-iepurasul/. Accessed 13 May 2014.
Gradinaru, Daniela. "Locuri pitoresti din Insulele Britanice ce merita vizitate." *Business 24*, 31 July 2012, www.business24.ro/vacanta/destinatii-turistice/locuri-pitoresti-din-insulele-britanice-ce-merita-vizitate-1516769. Accessed 13 May 2014.
Ligia. Comment to "Biblioteca – volumul II." *Viata mai simpla*, 22 January 2014, www.viatamaisimpla.blogspot.ie/2014/01/biblioteca-volumul-ii.html. Accessed 13 May 2014.
Luckyrock. Comment to "Cartea de miercuri – Povestea lui Peter Rabbit." *Idei pentru mămici si copii*, 12 January 2011, www.laurafrunza.com/2010/04/07/cartea-de-miercuri-povestea-lui-peter-rabbit/.
Melaroana. "Numaratoarea inversa pana la Craciun: ziua 5." *Viata, suflet, dragoste*, 20 December 2011, www.melaroana.wordpress.com/category/zi-de-zi/page/13/. Accessed 13 May 2014.

Pete, Mirela. "Ilustratori care imi plac: Beatrix Potter." *Mirela Pete. Blog*, March 2012, www.mirelapete.dexign.ro/2012/03/ilustratori-care-imi-plac-beatrix-potter/. Accessed 13 May 2014.

Popescu, Diana. "Povestea iepurasului Peter - Desene animate." *Pentru copii*, 9 February 2013, www.copiipovesti.blogspot.ie/2013/02/povestea-iepurasului-peter-desene.html#comment-form. Accessed 13 May 2014.

Sarbu, Daciana. "Lectii pentru fetita mea." *Green Report*, 14 February 2008, www.green-report.ro/daciana-sarbu-lectii-pentru-fetita-mea/. Accessed 13 May 2014.

Silving. Comment to "Ilustratori care îmi plac." *Mirela Pete. Blog*, 1 February 2013, www.mirelapete.dexign.ro/2012/03/ilustratori-care-imi-plac-beatrix-potter/.

"Trei iepurași de poveste." *Parinti.com*, www.parinti.com/Treiiepura537idepoveste-articol-3547.html. Accessed 16 August 2017.

Vida-Rațiu, Camelia. "The Tale of Peter Rabbit." *Mic atelier de creație*, 19 June 2013, www.micatelierdecreatie.ro/2013/06/the-tale-of-peter-rabbit.html. Accessed 13 May 2014.

Secondary Sources

Al-Yaqout, Ghada and Maria Nikolajeva. "Re-conceptualising Picturebook Theory in the Digital Age." *Barnelitterært Forskningstidsskrift*, vol. 6, no. 1, 2015, https://doi.org/10.3402/blft.v6.26971.

Bartow Jacobs, Katrina Emily. "The (Untold) Drama of the Turning Page: The Role of Page Breaks in Understanding Picture Books." *Children's Literature in Education*, vol. 47, 2016, pp. 357–373.

Barza, Vlad. "Boom-ul internetului in Romania: De la un inceput timid, la febra netcafe-urilor si viteze de top." *Hotnews*, 12 November 2013, www.economie.hotnews.ro/stiri-20_ani_internet-16011418-boom-internetului-romania-inceput-timid-febra-netcafe-urilor-viteze-top.htm. Accessed 22 March 2017.

Bruscini, Veronica. "'So I Shall Tell You a Story': The Subversive Voice in Beatrix Potter's Picture Books." *Honors Projects Overview*, Rhode Island College, 2008, www.digitalcommons.ric.edu/cgi/viewcontent.cgi?article=1028&context=honors_projects. Accessed 27 August 2014.

Carpenter, Humphrey. "Excessively Impertinent Bunnies: The Subversive Element in Beatrix Potter." *Children and Their Books: A Celebration of the Work of Iona and Peter Opie*, edited by G. Avery and J. Briggs. Oxford University Press, 1989, pp. 271–298.

Cocargeanu, Dana. "The Adventures of Peter Rabbit in Romania: Translation Challenges and Strategies." *International Research in Children's Literature*, vol. 7, no. 2, 2014, pp. 198–212.

———. "Off the Beaten Track in Romanian Translation Studies: Translations for Children. Visual Elements in the Romanian Translation of *The Tale of Jemima Puddle-Duck* by Beatrix Potter." *Constructions of Identity (VII)*, October 2013, Cluj-Napoca, Romania, edited by Adrian Radu, Casa Cărţii de Ştiinţă, 2014, pp. 74–88.

———. *Children's Literature Across Space and Time: The Challenges of Translating Beatrix Potter's Tales into Romanian*. Dissertation, Dublin City University, 2015, https://capitadiscovery.co.uk/dcu/items/727305?query=cocargeanu&resultsUri=items%3Fquery%3Dcocargeanu.

Cocargeanu, Dana, Eithne O'Connell and Áine McGillicuddy. "Beatrix Potter: A Hundred Years in Translation." *Children's Literature in Education*, vol. 47, no. 4, 2016, pp. 374–391.

Coitit-Godfrey, Janie. "Beatrix Potter in France." *Beatrix Potter Studies*, vol. 5, 1993, pp. 54–57.

Demourova, Nina. "Peter Rabbit in Russian". *Beatrix Potter Studies*, vol. 10, 2003, pp. 69–74.

Doonan, Jane. *Looking at Pictures in Picture Books*. Thimble Press, 1993.

Dresang, Eliza T. "Radical Change Revisited: Dynamic Digital Age Books for Youth." *Contemporary Issues in Technology and Teacher Education*, vol. 8, no. 3, 2008, pp. 294–304.

Flewitt, Rosie. "Reading Transformations". *Children's Literature: Approaches and Territories*, edited by Janet Maybin and Nicola J. Watson. Palgrave Macmillan and The Open University, 2009, pp. 352–366.

Genette, Gerard. *Paratexts: Thresholds of Interpretation*. Translated by Lewin, Jane E. Cambridge University Press, 1997.

Henkel, Ayoe Quist. "Exploring the Materiality of Literary Apps for Children." *Children's Literature in Education*, 2016, https://doi.org/10.1007/s10583-016-9301-7, n.p.

Higonnet, Margaret R. "The Playground of the Peritext." *Children's Literature Association Quarterly*, vol. 15, no. 2, 1990, pp. 47–49.

International Telecommunications Union. "Percentage of Individuals Using the Internet." n.d., www.itu.int/en/ITU-D/Statistics/Pages/stat/default.aspx. Accessed 22 March 2017.

Ippolito, Margherita. "The Relationship Between Text and Illustrations: Translating Beatrix Potter's Little Books into Italian." *Whose Story? Translating the Verbal and the Visual in Literature for Young Readers*, edited by Maria González Davies and Riitta Oittinen. Cambridge Scholars Publishing, 2008, pp. 85–96.

Joosen, Vanessa. "From *Breaktime* to *Postcards*: How Aidan Chambers Goes (Or Does Not Go) Dutch." *Children's Literature in Translation: Challenges and Strategies*, edited by Jan Van Coillie and Walter P. Verschueren. St. Jerome Publishing, 2006, pp. 61–78.

Ketola, Anne. "Peter Rabbit in the Garden of Terror: Patronizing the Reader in Picturebook Translation." *Bookbird: A Journal of International Children's Literature*, vol. 55, no. 3, 2017, pp. 12–19, https://doi.org/10.1353/bkb.2017.0034. Accessed 30 October 2017.

Lathey, Gillian. "Introduction." *The Translation of Children's Literature: A Reader*, edited by Gillian Lathey, Multilingual Matters, 2006, pp. 1–12.

McDonough Dolmaya, Julie. "Revision History: Translation Trends in Wikipedia." *Translation Studies*, vol. 8, no. 1, 2015, 16–34, https://doi.org/10.1080/14781700.2014.943279.

Mullins, Kate. "Crossing Boundaries and Forming Identity in Beatrix Potter's *The Tale of Peter Rabbit* and *The Tale of Benjamin Bunny*." *The Looking Glass: New Perspectives on Children's Literature*, vol. 13, no. 1, pp. 1–10.

Nikolajeva, Maria and Carole Scott. *How Picturebooks Work*. 2001. Routledge, 2006.

Nodelman, Perry. *Words About Pictures: The Narrative Art of Children's Picture Books*. University of Georgia Press, 1988.

Otsuki, Ruriko. "Beatrix Potter in Japan." *Beatrix Potter Studies*, vol. 5, 1993, pp. 19–28.

Pascua-Febles, Isabel. "Translating Cultural References. The Language of Young People in Literary Texts." *Children's Literature in Translation: Challenges and Strategies*, edited by Jan Van Coillie and Walter P. Verschueren. St. Jerome Publishing, 2006, pp. 111–121.

Pearson, Lucy and Peter Hunt. *Children's Literature: Texts, Contexts, Connections*. Pearson Education Limited, 2011.

Sale, Roger. *Fairy Tales and After: From Snow White to E. B. White*. Harvard University Press, 1978.

Scott, Carole. "Clothed in Nature or Nature Clothed: Dress as Metaphor in the Illustrations of Beatrix Potter and C. M. Barker." *Children's Literature*, vol. 22, 1994, pp. 70–89.

———. "Between Me and the World: Clothes as Mediator Between Self and Society in the Work of Beatrix Potter." *The Lion and the Unicorn*, vol. 16, no. 2, 1992, pp. 192–198.

Urba, Kestutis. "The Perils Peter Rabbit has Faced in Lithuania." *Beatrix Potter Studies*, vol. 10, 2003, pp. 79–84.

Ward, Mark. "Harry Potter and the German Pirates." *BBC News*, 1 September 2000, www.news.bbc.co.uk/2/hi/science/nature/906048.stm. Accessed 21 April 2017.

Willsher, Kim. "Harry Potter and the Boy Wizard in Paris." *The Guardian*, 8 August 2007, www.theguardian.com/world/2007/aug/08/france.harrypotter. Accessed 21 April 2017.

Yoshida, Shin-ichi. "A Case of Distortions." *Beatrix Potter Studies*, vol. 10, 2003, pp. 75–78.

Zafiu, Rodica. "Limba română contemporană: Stilistica." *Universitatea din București*, 28 January 2014, www.unibuc.ro/prof/zafiu_r/Limba_romana_contemporana_Stilistica.php. Accessed February 2014.

Dana Cocargeanu holds a Ph.D. in children's literature and translation studies from Dublin City University, Ireland. Her thesis, *Children's Literature across Space and Time*, focuses on the translation of Beatrix Potter's works for a contemporary Romanian audience. Her research interests include Beatrix Potter, picture books, ideology in (translated) children's books, and children's books in communist and post-communist countries. Her research on the translation of Beatrix Potter has been published in *International Research in Children's Literature* (2014) and *Children's Literature in Education* (2016).

Between Light and Dark: Brazilian Translations of Linguistically Marked Ethical Issues in Star Wars Transmedia Narratives for Children

Domingos Soares and Cybelle Saffa Soares

There has been an awakening. Have you felt it?
The dark side… and the light.

After Disney purchased Lucas Film LMT from George Lucas, Star Wars has made its way back from a far-away not-so-long-ago galaxy with a new filmic installment, *The Force Awakens* (Abrams). Also called *Episode VII*, the feature film not only provided a fresh, expanded perspective into the storyworld but also opened up a venue for a plethora of publications in Brazil, including some of the first materials for children. New and old books, comics, games, films, and series flooded bookshop shelves, online retailers, pay television channels, and movie theaters: transmediation meets Star Wars in Brazil. A catalyst for this initial excitement—*The Force Awakens* teaser trailer[1]—begins with an off-screen, deep, sinister voice uttering the words you read in the epigraph. The first massive insight the worldwide public has into this new era brings to fore the saga's

D. Soares · C. S. Soares (✉)
Universidade Federal de Santa Catarina, Florianópolis, Brazil

A. Kérchy and B. Sundmark (eds.), *Translating and Transmediating Children's Literature*, Critical Approaches to Children's Literature,
https://doi.org/10.1007/978-3-030-52527-9_13

all familiar light–dark side dualism—the two sides of the mystical *Force* from which heroes and villains draw their supernatural powers in the Star Wars universe.

Light and *dark* embody the clash between the opposing forces involved in the ethical dualism between right and wrong, good and evil. For children, who are learning about morals and ethics through narratives, Star Wars offers reference to act in the real world. In addition to being symbols that represent what should be taken as reference or avoided, *light* and *dark* are also words, resources inherent in vernacular language. Drawing on Brazilian Portuguese resources to translate these terms occasionally leads to forms of evaluation and intensification or mitigation of the values for which the source text terms stand. This last aspect is crucial to our analysis since having children as target audience justifies the didactic principle of teaching about morals through the representations of fictional characters and metaphysical entities, such as the Force.

In Brazil, the Disney era has introduced new translational options that broaden the range of available representations of the dark–light dualism. With the variety in publications come a diversity in translation options of some of the key terms in the narrative world, such as *light* and *dark*. As books, comics, films, and games are composed of diverse meaning-making elements, they are approached differently in translation. Therefore, the mediating role of translation to *re*present *light* and *dark* is limited by the medium-specific requirements each different medium poses. Star Wars is ultimately a story told in multiple media and in multiple texts that compose one fictional world. Due to the cross-textual nature of a transmedia narrative, the overall representation of the ethical dualism depends on the coordination in the production and translation of the installments that build the fictional world.

In the present chapter, we investigate the translation of the terms *light* and *dark* when they represent the ethical dualism between good and evil—be the terms alone or part of word strings—in Brazilian translations of the Star Wars transmedia franchise. In other words, how does the examined world built in the translations represent the ethical dualism embodied by the terms *light* and *dark*? It should be stressed that it is not our purpose to exhaust the translational decisions taken but to explore as wide a variety of translation options as we can. From a theoretical perspective, we abide by the basic premises devised by Descriptive Translation Studies. That is to say, our main goal in the present study is to describe translational behavior rather than to prescribe or evaluate translation quality, and to try to explain it based on a set of established variables (Chesterman). These variables are (1) target language, namely, the different resources in the languages involved; (2) reader, the specific needs of the child audience; (3) media, the multimodal affordances and technicalities in each medium; and (4) transmedia principles related to continuity and multiplicity (this last variable is not contemplated by Chesterman). We use these variables to investigate what might have influenced the translation decisions. How do the aspects related to language, media-specificity, target audience, and transmedia strategy interact and influence the translation decisions? Finally, given the didactic principle by which children's stories abide, we reflect on the way they reproduce the dualism addressed above. How could the translation options inform children's perception of good and evil?

Because of the vastness of these publications, it is unfeasible to investigate the integral franchise. Therefore, the selection of the corpus examined met the following criteria:

- Diversity in translated media: novels, audiovisual content, games, and comics (to observe how their medium-based specificity might influence translational decision-making)
- Theme: particular stories whose theme and plot focus on the mystical Force and the role of the characters who act as agents of the *dark* and the *light*, in detriment to the stories that address themes like politics or war
- Age group: materials aimed at children in the source context or rated as free for all audiences in Brazil
- Period: materials first published or relaunched in 2015 and 2016, after Disney's purchase of the franchise
- Availability: materials that compose our private collection (put together throughout the years as Star Wars fans), incremented by borrowings from other collectors and from the children's library Barca dos Livros (in Florianópolis), as well as content on Netflix.

Owing to space constraints, we reference the selected materials throughout the paper instead of listing them here. For now, it suffices to mention that not all materials investigated are explored here, only the ones containing the more emblematic translation options. These include novels and a comic book targeted at children from nine to twelve years old (according to the Disney Books website),[2] television series rated as free for all audiences by Brazilian authority,[3] and a game not advisable for those younger than twelve years old. Before we move on to analyze the translations, we explore some of the fundamental transmedia principles of interest to the present study.

Translating a Transmedia Narrative

According to Henry Jenkins, "transmedia storytelling is the art of world making" (*Convergence Culture* 21). The various installments of a storyworld realized across distinct media platforms supersede the fictional elements of each particular installment, enhancing the storyworld. In combination, they comprise a network of interconnected stories that function in a complementary manner and ultimately create a composite narrative whole, a fictional universe. About imaginary worlds, such as Star Wars, Mark Wolf contends that—in addition to the basic elements of space, time, and characters—underlying structures comprise the world itself. Among them, language "contains a culture's worldview embedded within it" (154), regulates expressive forms, and "gives communicable form to the way in which the members of a culture collectively conceptualize their world" (155). The heroic Jedi Knights and villainous Sith Lords are inhabitants of a universe where *light* and *dark* are defining features of cultural reality and linguistic expression of their ethics. Nevertheless *light* and *dark* are also English language words. Translation is in charge of the *re*making of a world. Translation decisions might encode diverse representations of the dualism—with special repercussions on the child audience who is learning the boundaries of good and evil, and learning to distinguish fantasy from reality.

According to Jenkins, the fictional elements that make up the fictional world "require a high level of coordination and creative control," and "all of the pieces have to cohere into a consistent narrative or world" to ensure continuity ("Transmedia 202"). In contrast, transmedia projects might also give room to some degree of multiplicity by allowing alternative retellings or narrative elements that systematically differ from one another. The continuity versus multiplicity principle might be a source of confusion to the child audience that more easily follows linear narratives with a beginning, climax, and an end, turning this particularity into a challenge to translation. Multiple versions of a story or narrative element, such as alternative translations for light and dark in different contexts within the transmedia narrative, might also provide the opportunity for a broader comprehension of the ethics they symbolize. Additionally, the crossing of media platforms also entails giving different (and complementary) insight into the fictional world because each media has its specificities and meaning-making potentials, something Gunther Kress calls affordances: "different modes offer different potentials for making meaning" (79). With them come different translation methods (i.e., dubbing and subtitling in audiovisual materials) and technicalities (e.g., space constraints in comics' translations and subtitling and game localization). Some of the challenges these technicalities might impose are also an issue when it comes to translating for children, given their in-development literacy skills and linguistic knowledge. These aspects inform the translational decisions that are investigated next.

Translating the *Light*

In Western tradition, *light* is "traditionally linked with goodness, life, knowledge, truth, fame, and hope" (Ferber 115). It is a symbol of moral and intellectual superiority. From a purely physical perspective, it is but electromagnetic radiation. Used in imaginary environments, symbolic appropriation from "real world" cultural inventory relies on the iconic depiction that goes along with the uses of the term, and Star Wars does not disappoint us. As Jedi Master, Ben Kenobi helps us understand, "the light side of the Force, the way of the Jedi, aligns with selflessness, enlightenment, mercy, and compassion" (Bracken 104). The spaceship that takes us to the Star Wars storyworld is fueled by Western tradition: Even being light-years apart, the *light* remains the same. However, not all transitions can be taken that lightly, especially when it comes to translation.

Heirs of the same cultural tradition, both symbolic and literal meaning match the word *luz* in Brazilian Portuguese. No wonder this is a very common translation option. It is convenient as a translation solution for a series of other reasons. First, due to its frequency of use, *luz* is a lexical item that enters the inventory of a child early in life, which contributes for a readable translated text. Second, from the technical perspective, translating comics, localizing games, and subtitling audiovisual content can be challenging on account of the limited space available (Díaz-Cintas and Remael 9). Dubbing requires translation solutions that synchronize with onscreen actor's lip movement (Linde and Kay 2). For one, *luz* is short enough to dismiss space limitation issues; it is also a one-syllable word that begins with the same consonant sound. The harmony in meaning, audience needs, and technical aspects might make translating *light* into *luz* unproblematic enough to use it consistently and, as a result, guarantee transmedia continuity across several installments. This translation option can be observed, for example, in the novel *Cuidado com o Lado Sombrio da Força*:

He has one last thing to say that might help to tip the balance toward the **light**.	Ele tem uma última coisa a dizer que pode ajudar a fazer a balança pender em direção à **luz**.

A facet of the dual Force, *light* in the usual string *light side* can be interpreted as an adjective. Unlike in the English language, changing word position in a sentence is often not enough to turn a noun into an adjective in Brazilian Portuguese. Some degree of morphological adjustment is commonly the alternative used to translate *light side* in the analyzed texts. As the idea of light is heavily loaded with symbolic and cultural meanings in Western tradition, morphosyntactic shifts involving the term might interfere with the representation of the ethical dualism because they can generate translation options that could evoke diverse metaphorical associations and evaluative portrayals in the target context. One of the translation options for *light side* identified is *lado iluminado* ("enlightened side" or "illuminated side"). The adjective derives from the same root word as the English cognate *illumination*. From this perspective, those characters associated with the *light side* are thus imbued with knowledge of the nature of reality and plain spiritual awareness, as in the following extract from the animated series *The Clone Wars* (O'Connell):

The Father keeps a fragile balance between his daughter, who allies with the **light side**, and the Son, who drifts ever closer to the dark.	O pai mantém o frágil equilíbrio entre sua filha, que segue o **lado iluminado**, e seu filho, que prefere o lado sombrio.

In the excerpt from the television show *The Clone Wars* (Season 3, Episode 16), the so-called daughter[4] is a near spirit-like being: her body literally emits light. Despite having adopted a longer translation solution, which by the way is the same in dubbing and subtitling, length in characters and syllable count is compensated by abridging other parts of the sentence. Concurrence between visual design, a conceptual representation of the character and translation option might also offer the child audience an opportunity to broaden their linguistic repertoire. Under these grounds, adopting translation options that highlight spiritual attributes is a way to evaluate this particular character positively and, by extension, other characters and ideas that share the same principles. To the dualism embodied by *light* and *dark*, this implies *improving the good*.

Another form of evaluative translations of *light side*, but this time less metaphorical, has been identified in the *Star Wars: Galaxy of Heroes* game for portable devices (Bracken). The player can fight either light- or dark-side battles. In Brazil, the former is rendered as *lado bom* ("good side") (Fig. 1).

At the bottom left side of the picture, you read "Batalhas do Lado Bom" ("Light Side Battles"). According to O'Hagan and Mangiron,

Fig. 1 Screenshot from the game *Star Wars: Galaxy of Heroes*

"text in the UI [user interface] is subject to strict space limitations, particularly in menu screens, which usually contain different types of information" (122). In this particular case, *bom* ("good") seems to have been chosen as the translation option owing to space constraints. Even though we have considered the Brazilian Portuguese noun *luz* as a convenient translation option based on space constraints, the present case demands an adjective. Thus, using *luz* would require some degree of morphosyntactic adjustment, which in turn would make it longer and possibly unfit for the game's user interface.

One of the distinctive features of games is that players are not simply told a story; they impersonate the characters and actively live in the fictional world. In this particular case, players can only initially take the role of the good guys. Only after making some progress in the game will the player be able to fight dark-side battles, in which case they unescapably have light siders as opponents. The underlying metaphorical association between the *light* and superior qualities depicted in a text can potentially reflect not simply the legitimacy of the *light* as a metaphor but the very qualities that are assumed to stand for the *light* in a given sociohistorical context. *Good*, on the other hand, can be read as a priory evaluation that sets the standards for desirable conduct. In other words, by assuming that someone is good, their behavior and attitudes become the reference against which appropriateness can be measured. That can be particularly problematic in games that involve armed conflict and war, even if they are rated inappropriate for twelve-year-olds or younger, because it seems as though violent acts are being justified in the intrinsic righteousness of the "good side" they embody in gameplay. Accordingly, the opposing side might appear to be inescapably evil and deserving of punishment, which good-side players could uncritically inflict. In the same vein, players joining the "dark side" could openly oppose the "good side" and impersonate the "evil side," or even the "black side" (evoking rather politically incorrect implications in English). Next, we discuss some of the more significant translation options of *dark (side)* we encountered.

Translating the *Dark*

According to Michael Ferber's *A Dictionary of Literary Symbols*, *dark* in Western cultural tradition is initially the utter absence of light: "[w]hen all was darkness, the first thing God created was light (Gen. 1.3)." From this perspective, the dark is a precondition of existence, which has come about upon the creation of the light. Even though it is traditionally associated with "evil, death, ignorance, falsehood, oblivion, and despair" (Ferber 115), Cirlot reminds us that "[t]he dualism of light/darkness does not arise as a symbolic formula of morality until primordial darkness has been split up into light and dark" (76). That is to say, the dark being considered "evil" is a result of the light ontology defined in opposition to it, not an intrinsic feature in the dark, which by definition lacks ontology of its own: the absence cannot be; nothing can never be something. The traditional association the dark elicits is a result of sociohistorical representation in context across the centuries. In Star Wars, let us once again draw on Jedi Master Obi-Wan Kenobi's teachings: the dark side "deals in hatred, fear, anger, jealousy" (Bracken 104). As stated previously, target audience needs, technical translation imperatives, transmedia principles, and contrastive linguistic aspects inform the translation decisions and the way they represent the light–dark ethical dualism.

Differently from some of the translations of *light* (and its *luz* translated counterpart), none of the solutions adopted for *dark* match the literal and symbolical meanings and similarity in morphology so conveniently. That is, each translation provokes a different effect that highlights or cancels some feature of the *dark* mythical entity. The subtitled translation solution in the example below (from *Clone Wars,* Season 3 Episode 12), *trevas*, is the conventional equivalent of *dark* in "dark ages" ("idade das trevas").

I will teach you the ways of the **dark side**.	Vou ensinar o caminho das **trevas**.

Besides the ominous associations it might arouse, according to *Houaiss Dictionary*, *trevas* also stands for "the total absence of light." This definition resonates with the idea that the *dark* is not something in itself. This particular rendition seems to reflect such understanding of the *dark* since it rejects the *side* in "dark side." Shorter subtitles make for optimal comprehension of the filmic text because viewers need to have the opportunity to read both verbal and visual information on the screen. That is particularly true for younger audiences who are developing their reading skills.

Another significant use of *trevas* is the translation option to characterize the representatives of the dark side in the storyworld of the *Dark Lords of the Sith*:

But he turned to the **dark side** and killed his master. He is now a **Dark Lord** of the Sith.	Mas ele se voltou para o **lado sombrio** e matou o mestre. Agora ele é o oposto de um mestre Jedi. Ele é *o* **Senhor das Trevas** dos Sith.

As it is used, the idea of *dark* is that of binary opposition embedded in moving from one *side* to the other. In the story, a character "*turned* to the dark side." In this context, the *dark* is not the mere absence of light but the opposite of light. Moreover, the translated excerpt is significantly longer than the source one. There is a full sentence addition (back-translated as "Now he is the opposite of a Jedi Master") with the intention to justify the violent connotation of the verb *kill* by foregrounding the appositive condition of villains and heroes. Additions of this sort are often unfeasible in multimodal context-given space limitations.

Moreover, in a multimodal context with meaning-making resources other than verbal language, the very depiction of Jedi and Sith (such as in the game discussed previously) might be considered as sufficient to account for their opposition. Black masks or hissing voices from under hooded, shapeless faces produce strong pictorial impact. In monomodal environments where verbal language is the only meaning-making resource, the power of definition is as sharp as lightsabers. The *Dark Lords* are the *Senhores das Trevas* in translation. The villains in the story are comparable to those sinister beings that permeate popular imaginary, such as "filho das trevas" ("son of darkness") or "príncipe das trevas" ("dark prince"), common epithets of popular characters such as Dracula or the Devil. As far as our exploration went, *trevas* is not the standard translation solution in all occurrences of the term *dark*. Using alternative options to translate *dark* mitigates the dread evoked by a more ominous *trevas*, even at the expense of weakening transmedia consistency by maintaining the same translation option.

In the same excerpt addressed above, we observe the phrase "dark side" being translated as *lado sombrio* ("shadowy side"). Unlike *trevas*, which can be understood as the absence of light, a shadow is the result of light being intercepted by an object. The amorphous uncanny darkness in *trevas* gains contours and shapes as a shadow. According to Cirlot, "[a]s the Sun is the light of the spirit, so *shadow* is the negative 'double' of the body, or the image of its evil and base side" (290). In *shadowy side*, the ethical issue encoded in the symbol remains, but its physical facet calls for more literal iconic associations. To this end, the feeling of fear that the uncanniness in *trevas* might inspire appears to be mitigated in *sombrio* because shadows are close to the child's world of experience. Additionally, this translation option also eschews the associations with sinister characters that *trevas* has. From a transmedia perspective, using different translation solutions for the same term (*sombrio* for "*dark* side" and *trevas* for "*Dark* Lord") might lead to continuity problems, especially when they are used adjacent to each other. In this particular case, continuity might not develop into an issue because the adjective *dark* is attributed to different nouns, namely, a facet of the mystical entity ("dark side") and a category of characters within the storyworld ("Dark Lords").

There seems to be some degree of coordination driven toward producing consistent translations across texts and media. In the case below, *lado sombrio* is also the translated option. However, the source text counterpart reads "dark path," such as the following extract from the comic book *Star Wars: O Império Contra-ataca*:

If once you start down the **dark path**, forever will it dominate your destiny. Consume you it will, as it did Obi-wan's apprentice.	Se começar a trilhar o **lado sombrio**, pra sempre seu destino ele dominará, como fez com o aprendiz de Obi-wan.

Once again, available space is an issue that constrains translation options. As Grun and Dulleroup put it, translation of comics is "limited spatially in that translations must fit into balloons or panels, and in that they have a specific objective" (198). "Dark path," which could be translated as the longer *caminho sombrio*, is rendered as not only a shorter version but also a solution that maintains consistency with translation options in other materials.

In the first Star Wars film translations, the "dark side" was translated as *lado negro*. The word *negro* can refer to the color black (also *preto* in Brazilian Portuguese), such as in *ouro negro* ("black gold"), but it is often used in more evaluative contexts, as in *mercado negro* ("black market"), *humor negro* ("black humor"). *Negro* is also the word used to refer to race: *negro* ("black man/black person"), *negra* ("black woman"). Accordingly, *Lado negro* can be interpreted as "black person's side." Even though the word *negro* is not often considered prejudicial in itself, coupling race with questionable ethics can be seen as problematic in a country where slavery is a part of history. *Lado negro* has been removed from the short animated series *Yoda Chronicles* (Hegner) in recent retranslations, but it still seems to have its echoes. That can be observed in the subtitled version of the first episode:

In your face, **dark side**.	Toma, **Lado Negro**.

Despite the negative associations this translation option can inspire, one of the translators in charge of the most recent retranslations of the Star Wars feature films has stated that the decision to use *lado sombrio* in the films has been made considering translation appropriateness rather than political correctness (Freire). The detrimental perception of *negro* might not be controlled by the agents involved in the translation process. Nevertheless, this sort of representation could result in associations that might perpetuate prejudice because children, who are developing their literacy skills, are still learning to understand ambiguity and the distinction between reality and fantasy. The dubbed rendition (from *Lego Star Wars: The Yoda Chronicles*) of the same dialogue line loosely back-translates as "you evil thing."

In your face, **dark side**.	Na tua cara, **ô bicho do mal'**.

In cartoons, dubbing lip synchronicity is occasionally dismissible because realism in character design is not an issue. This is particularly true in the present case because characters simulate toys, and they do not represent real evil humans; thus, translators can take more liberties with their lexical choices. Margery Hourihan states that "characters are distanced from the audience by their dehumanization" (53). In the case at hand, the hero (in the center of the image below from *Lego Star Wars: The Yoda Chronicles*) teases the villain upon his victory (Fig. 2).

The translation emphasizes the vocative function in the source text—personalizing the opponent, who is branded outright as *evil* by the hero. Additionally, using informal language reinforces the propensity of the audience to relate even closer to the hero. The hero is cool; the villain is evil. This evaluative stance on either end of the dualism reproduces an over simplistic worldview, but one that can still be considered strategic in developing the very sense of right and wrong in young children and one

Fig. 2 Scene from *Lego Star Wars: The Yoda Chronicles*

that is reinforced by the need of having any danger resolved in the end. Ann Trousdale argues that “the punishment of the villain in the tales does not seem to have a pathological effect upon children” (77). Instead, it has a positive effect on their future adult life, for it helps them understand the morality constraints of good versus evil.

To sum up, in the case of *light*, some of the translations share etymological, referential, and symbolic similarities. However, given the multimodal aspects of audiovisual translation and game localization, other translation options could be interpreted as rather evaluative and less metaphorical, such as “good side,” or they could stress spiritual superiority, such as “illuminated side” (both for “light side”). Conversely, *dark* draws on slightly different metaphors and conceptualizations. Space constraints and multimodal affordances highlight both the non-emphasis on the dualism by omitting the word *side* in translating “dark side” and the sinister, powerful perception of the *dark* characters, in *Senhor das Trevas* (“Dark Lord”). Differently, *lado sombrio* (“dark side”) evokes uncannier associations and is considered the correct rendering by some in the translator community. Away from symbolic meanings, *lado negro* (“black side”) can be considered a sensitive rendering because, from a certain perspective, it can be associated with race and questionable ethics.

In this particular case, we suppose there is an echo of the original feature film translated back in the 1980s.

Conclusion

From start, Star Wars' creator George Lucas strategically adopted the dualism between light and dark as a vital organizing principle of his fictional universe, meant to communicate an ethically invested moralizing message in line with the high fantasy heroic quest narratives. In Lucas's words, "the Force evolved out of various developments of character and plot. I wanted a concept of religion based on the premise that there is a God and there is good and evil … I believe in God and I believe in right and wrong" (quoted in Eberl and Decker 196). The terms *light* and *dark* embody this worldview in Star Wars. They not only stand for entities in the metaphysics of the fictional world but also contribute to the very characterization of the world. These words create the world; their translations recreate the world.

As remarked earlier, translating a transmedia story for children comprises aspects such as inherent differences in the languages involved, technical specificities in each media, target audience requirements, and transmedia principles. Even though these factors pose a challenge to translation, they do not determine the chosen translation option; they rather delimit other possibly available options. In the case at hand, we argue that the variety of factors influencing the translations have given rise to such a diversity of translation solutions. From a transmedia storytelling perspective, diversity in an element that contributes to creating the fictional world can render it inconsistent, one that does not maintain continuity across the several installments. According to Jenkins (Transmedia 202), the principle of continuity "requires a high level of coordination and creative control and that all of the pieces have to cohere into a consistent narrative or world." The multiple translations identified could also be a sign of the varying degrees or lack of coordination to which the translations were submitted. The multiplicity in the translations identified can additionally be a result of translators' usual freedom to act. As Shavit explains, "the translator of children's literature can permit himself great liberties regarding the text" because of the low status of children's literature (112). However, the liberty mentioned by Shavit is subject to the translator's and the target culture's notions of childhood, including what is considered beneficial and appropriate for children to know and learn.

Accordingly, little or lack of coordination, in this particular case, allows for the target culture's conceptions about childhood to step in instead of simply attempting to reconstruct the terms used by George Lucas and, with the terms, his own worldview.

Scattered as the planets in that far-away galaxy, the several installments of Star Wars require curious pilots to explore them. As the pilots get to know the galaxy, they realize there is more and more to explore. Traveling in a transmedia storyworld involves creating and reworking a network of relations. They do not simply travel by automatic pilot: "young people make these stories their own through their active imaginations" (Herr-Stephenson, Alper, Reilly and Jenkins). Being a pop culture icon developed under the premise that there is good and evil, Star Wars becomes one reference for the representation of ethical issues in society. The multiplicity in the translations of *light* and *dark* offer the child audience raw material from which to build their own conceptions of the ethical dualism. The mediation of translation offers relevant contributions by providing the audience with ample representations of the ethical dualism, but the child audience share efforts in constructing their own worldview.

Notes

1. https://www.youtube.com/watch?v=zdDi7pf26i4.
2. http://books.disney.com/book/beware-the-power-of-the-dark-side/ http://books.disney.com/book/so-you-want-to-be-a-jedi/http://books.disney.com/book/star-wars-the-original-trilogy/.
3. http://portal.mj.gov.br/ClassificacaoIndicativa/jsps/DadosObraForm.do?select_action=&tbobra_codigo=46640http://portal.mj.gov.br/ClassificacaoIndicativa/jsps/DadosObraForm.do?select_action=&tbobra_codigo=14917.
4. Picture: http://www.starwars.com/databank/daughter.

Works Cited

Abrams, J. J. *Star Wars: The Force Awakens*. Walt Disney Studios Motion Pictures, 2015.

Angleberger, Tom. *Star Wars: The Return of Jedi. Beware of the Power of the Dark side*. Disney LucasFilm Press, 2015.

Bracken, Alexandra. *Star Wars: A New Hope. The Princess, the Scoundrel, and the Farm Boy*. Disney Lucasfilm Press, 2015.

Eberl, Jason and Kevin Decker. *The Ultimate Star Wars and Philosophy: You Must Unlearn What You Have Learned*. John Wiley & Sons, 2015.

Gidwitz, Adam. *Star Wars: The Empire Strikes Back. So You Want to Be a Jedi?* Disney Lucasfilm Press, 2015.

Hattner, Álvaro. *Star Wars: uma nova esperança, a princesa, o cafajeste, e o garoto da fazenda*. Seguinte, 2015.

Hegner, Michael. "The Phantom Clone." *Lego Star Wars: The Yoda Chronicles S01 E01*. Netflix, 2013.

Muniz, Maurício. *Star Wars: o retorno de Jedi, cuidado com o lado sombrio da força!*. Seguinte, 2015.

O'Connell, Brian Kalin. "Altar of Mortis." *The Clone Wars S03 E16*. Netflix.

Volpe, Giancarlo. "Nightsisters." *The Clone Wars S03 E12*. Netflix.

Secondary Sources

Chesterman, Andrew. "Empirical Research Methods in Translation Studies." *Erikoiskielet Ja Käännösteoria*, vol. 27, VAKKI-symposiumi XX, 2000, pp. 9–22.

Cirlot, Juan Eduardo. *A Dictionary of Symbols*. Courier Corporation, 2002.

Díaz-Cintas, J. and Aline Remael. *Audiovisual Translation: Subtitling*, 2014.

Ferber, Michael. *A Dictionary of Literary Symbols*. 2nd ed., Cambridge University Press, 2007.

Freire, Jorge. *Lado Luminoso Da Força, Lado Sombrio Da Força E O Lado Politicamente Correto | Precisa de Tudo Isso?* Retrieved February 2018, http://nerdpai.com/lado-luminoso-da-forca-lado-sombrio-da-forca-e-o-lado-politicamente-correto-precisa-de-tudo-isso/.

Herr-Stephenson, Becky, et al. "T is for Transmedia: Learning Through Transmedia Play." *Los Angeles and New York: USC Annenberg Innovation Lab and The Joan Ganz Cooney Center at Sesame Workshop*. 2015. Retrieved February 2018, http://www.joanganzcooneycenter.org/wp-content/uploads/2013/03/t_is_for_transmedia.pdf.

Hourihan, Margery. *Deconstructing the Hero: Literary Theory and Children's Literature*. Psychology Press, 1997.

Jenkins, Henry. *Convergence Culture: Where Old and New Media Collide*. NYU Press, 2006.

———. "Transmedia 202: Further Reflections." *Confessions of an Aca-Fan*. 2011. Retrieved February 2018, http://henryjenkins.org/blog/2011/08/defining_transmedia_further_re.html.

Kress, Gunther. *Multimodality: A Social Semiotic Approach to Contemporary Communication*. Routledge, 2010.

Linde, Zoé de, and Neil Kay. *The Semiotics of Subtilting*. Routledge, 2014.

O'Hagan, Minako, and Carmen Mangiron. *Game Localization*. Vol. 106, John Benjamins Publishing Company, 2013, https://doi.org/10.1075/btl.106.
Scolari, Carlos Alberto. *Narrativas Transmedia: Cuando Todos Los Medios Cuentan*. Deusto, 2013.
Shavit, Zohar. *Poetics of Children's Literature*. University of Georgia Press, 2009.
Trousdale, Ann. "Who's Afraid of the Big, Bad Wolf?" *Children's Literature in Education*, vol. 20, no. 2, 1989, pp. 69–79.
Wolf, Mark J. P. *Building Imaginary Worlds: The Theory and History of Subcreation*. Routledge, 2014.

Intergenerational Transmissions

A Thousand and One Voices of *Where the Wild Things Are*: Translations and Transmediations

Annalisa Sezzi

This chapter sets out to explore the different voices that can be heard in the first Italian translation and the more recent retranslation of a classic picturebook: *Where the Wild Things Are* by Maurice Sendak.

Picturebooks derive their meaning from the relationship between words and illustrations. Lawrence Sipe defines this interrelation as "synergistic," and shows how the reader's oscillation between the verbal and visual material can be seen as a form of "transmediation". However, this interpretative process turns out to be more complex as picturebooks come to life through the adults' reading performance. Thus, the chorus of the discursive presences detectable in children's texts and in their translations (O'Sullivan, *Comparative Children's Literature*), such as the voice of the narrator and of the translator, is joined by the voice of the adult reading aloud.

Since retranslations act as sounding boards for both textual and contextual voices (Alvstad and Assis Rosa), the aim of this analysis is to identify the changes of the voices in the translation (in 1969) and retranslation (in 2018) of Sendak's *chef d'œuvre*.

A. Sezzi (✉)
University of Modena and Reggio Emilia, Modena, Italy

A. Kérchy and B. Sundmark (eds.), *Translating and Transmediating Children's Literature*, Critical Approaches to Children's Literature,
https://doi.org/10.1007/978-3-030-52527-9_14

Voices, Retranslations, and Picturebooks

Voices in translation and retranslation are a complex phenomenon. The focus of this paper is mainly on textual voices, that is, the discursive presences that are "audible" within the translated text itself: the voice of the narrator, the different voices of characters, and "the translator's textually manifested voice" (Alvstad and Assis Rosa 3). However, "contextual voices" inherent to the translating process will not be disregarded.

In order to detect these voices, O'Sullivan (in *Comparative Children's Literature*) offers an effective theoretical framework to identify the translator's voice, based on the amendments of Seymour Chatman's scheme of narrative communication suggested by Giuliana Schiavi and by Theo Hermans so as to embrace translated texts. Specifically, Schiavi argues that the translator's voice can be detected in translations. It establishes a new relationship between the readers and the translated text that would be otherwise impossible because of the different language, different *Weltanshauung*, and cultural background of the author and of the target readers. This voice only partly overlaps with the author's—disclosing its presence, for example, at a metalinguistic level, in paratexts and footnotes (Hermans). According to O'Sullivan, who applies the model to children's literature, the voice of the translator can be also identified at the level of the narration when it does not replicate the voice of the narrator of the original. This other voice is termed by O'Sullivan "the voice of the narrator of the translation" (*Comparative* 109).

In fact, during the process of translation, the translator becomes the real reader of the source text given his/her competence in both the source and target language and culture. He/she is therefore able to recognize the implied reader of the target text and assume its role with the aim of transferring the communication for the new readers. In this way, he/she creates an extra-agency, the "implied translator," who envisages a new target implied reader. This is of key importance in children's literature, characterized by an asymmetrical communication and strongly influenced by the idea of the child prevailing in the target culture.

Indeed, the translator might have an idea of the reading child, molded by the predominant and culture-specific notion of childhood, that can be dissimilar from the author's supposed child audience—hence constructing a child-implied reader that is different from the child-implied reader of the source text (O'Sullivan *Comparative Children's Literature* 110). Similarly, the possible discrepancies between the implied readers of the source text and of the target text convey different ideas of the presupposed child.

Yet, the translation pact (Alvstad) makes the readers accept that in the translation there are the author's words and not the translator's. This is true especially in the case in which the intended readership is children. Furthermore, the construct of the implied reader can be questioned since "the implied author will be reconstructed by readers in very much the same way regardless of whether they read a translation or a non-translated text" (Alvstad 275). Nonetheless, the notion of the child is revealed by the translator's voice, and it might vary between the source text and the target text as well as between the translation and the subsequent retranslations (Oittinen, *Translating*; O'Sullivan, *Comparative*; Douglas and Cabaret; Lathey). So does the image of the adult reader reading the picturebook aloud. A case in point is the translation and the retranslation of Maurice Sendak's *Where the Wild Things Are* as the analysis of some of its fundamental double-spreads reveals.

Where the Wild Things Are in Italy

The first Italian publication of *Where the Wild Things Are* dates back to 1969. It coincides with the introduction of the modern picturebook in Italy, which was a breath of fresh air for the national children's literature, visually too stereotyped. Brunari's work of the 1940s was the only noteworthy exception to this aesthetic platitude (Pallottino 343)—made of an illustrated universe of children with "chubby rosy cheeks, big eyes and long eyelashes" (Personal interview with Rosellina Archinto 5/5/2007). Rosellina Archinto and her publishing house, Emme Edizioni, carried out this sort of "Copernican revolution" (Fochesato 17).

After having seen *Little Blue and Little Yellow* in the United States, Rosellina Archinto realized how Italian children's books were behind the times. This was the impetus behind the foundation of her own publishing house in Milan in 1966; she wanted to create high-quality children's books. Her publishing policy acted like a powerful cultural magnet, attracting the most important authors and illustrators of the time, both from Italy and from abroad. Archinto produced and translated more than three hundred picturebooks, including works by Enzo and Iela Mari, Eric Carle, and Tomi Ungerer. Unfortunately, Emme Edizioni had to close in the 1980s since Italy was not ready for its path-breaking catalog.

The first translation of *Where the Wild Things Are* has to be seen against this fertile and innovative background. It is not by chance that Rosellina Archinto asked an Italian writer and poet, Antonio Porta, to translate it. After the closing of Emme Edizioni, *Where the Wild Things Are* was re-published in 1999 (using Emme's first translation) by Babalibri, a publishing house working in partnership with the French École des loisirs and founded by Rosellina Archinto herself and her daughter Francesca. This edition, titled *Nel paese dei mostri selvaggi*, was published until 2012, when Sendak died.

In fact, the author's death is a watershed in the Italian publication history of Sendak's work. The copyrights were negotiated again and were obtained by another Italian publishing house, Adelphi. A retranslation was finally published in January 2018.

Controversies among experts and non-experts of children's literature surrounding some of the new translation choices characterized the aftermath of the publication of the retranslation. One of the most eagerly anticipated events of the 55th edition of the Bologna Children's Book Fair for the Italian audience was in fact a conference centered on Sendak's *Where the Wild Things Are* or, rather, on the two current Italian editions of the Sendakian *chef d'œuvre*. Many of the agents of its publishing process were present. It was held on March 28, 2018, with the title "Celebrating Maurice Sendak" and gathered together Francesca Archinto, the former publisher of the first Italian edition; Anna Castagnoli, an Italian author and illustrator; Giovanni Nucci, an Italian author and the conference moderator; Lisa Topi, the translator of the Adelphi Italian edition; and Matteo Codignola, editor and translator of the Adelphi publishing house. They shed light on the picturebook itself and also on some of the new translation strategies, making many of the "contextual voices" (Alvstad and Assis Rosa) of retranslation meet face-to-face. These "contextual voices" also expressed themselves in different websites and social media, explaining and justifying many choices because of people's comparison with the previous one.

The Italian Translation and Retranslation of *Where the Wild Things Are*

The differences between the first and the second Italian translation of *Where the Wild Things Are* are many. The second translator, Lisa Topi, briefly but explicitly comments on her translation in an Italian online journal on translation, confessing her worries when she started translating the picturebook because "generations of readers had known the founding book of the canon of children's literature through Porta's translation" ("Ritorno"). Against the burden of the first translation made by a renowned figure, she says she opted for returning to the "naked and living force" of the text. Her translation project is thus in line with Antoine Berman's "retranslation hypothesis" (1990) that wants the first translation to be more target-oriented than the retranslations. In order to achieve this goal, Topi observes that Porta's translation—being someway "didactic" because of its prosodic, indirect, diluted style and the use of diminutives—needed to be polished and synthetized. Again, in accordance with the "anxiety of influence" (Koskinen and Paloposki), the second translator remarks on some of the limits of the first translation, emphasizing the process of research at the basis of her work. In the following paragraph, the different translation strategies will be examined to see if the polemical attitudes are justified.

Translating the Title, or the Anxiety of Influence

As underlined by Nières-Chevrel ("Les albums de Maurice Sendak" 70), one of the first problems encountered when translating *Where the Wild Things Are* is the translation of the title (Desarthe 78). There are some intratextual relations that need to be preserved. Indeed, Sendak's title is strictly connected to the epithet "Wild Thing!" uttered by Max's mother at the beginning of the book (Nières-Chevrel, "Les albums" 70) as well as to the words portraying Max on his way "to where the wild things are" (Oittinen et al. 160). Crucially, the major challenges the translator has to take on are posed by two apparently simple and highly frequent words: the adjective "wild" and the noun "things." Analyzing the different solutions in Spanish, Finnish, German, Italian, and Swedish, Oittinen et al. underline that the first Italian translation, the Finnish translation, and the Spanish translation opt for "monsters" (159). This is in line with Nières-Chevrel ("Les albums" 70–71): Generally, the European translations of the title of Sendak's picturebook can be classified into two groups, with the Romance languages revolving around the noun "monster" and the Germanic languages around the adjective "wild."

Despite having a direct Italian equivalent, the adjective "wild" covers a wider semantic field than the Italian "selvaggio"—as pointed out by

Nières-Chevrel for the French term "sauvage" ("Les albums" 70)—evoking Jack London's *The Call of the Wild* and the translation problems it poses (Alunni). "Things" is as much problematic as "wild." Obviously, it does not refer to an entity as "object" but refers to an "être inqualifiable," an "unqualifiable being," as the characters of children's stories usually are (Nières-Chevrel, "Les albums" 70). Thence, it is open to many other diverse interpretations, such as the "child's negative emotions," often difficult to name (Oittinen et al. 160). Oittinen et al. point out that the use of "mostri" in *Il paese dei mostri selvaggi* ("In the land of the wild monsters") implies the attachment of a label to Sendak's things, thereby reducing the unlimited interpretative possibilities of "things." Yet, as Lisa Topi explains, she decided to keep the title of the first translation because any other alternative could not satisfactorily convey the "misleading neutrality" of the English word and be immediately remembered. In point of fact, the corresponding generic and all-embracing Italian term "cosa" is rarely used for animated beings, to the point that its use would have probably been deceptive and not as forceful as Porta's "mostri."

Taking into the linguistic constraints, Porta's "mostri" is convincing for a twofold reason. Any analyses of the title cannot transcend the interaction of the verbal and visual. In this case, the title should be seen in relation to the cover illustration. Sendak's cover image does not depict the hero, as cover illustrations customarily do (Marcus 12). Equally surprisingly, there is one wild thing represented; this might destroy "the suspense of an enticing title by featuring the setting or [and] the antagonist on the cover" (Nikolajeva and Scott 246) or otherwise enhance it (Nodelman 50). In fact, the cover shows Max's boat sailing on the left and a wild thing on the right. Thus, Max's trip is portrayed together with one of the "things," which acquire a shape and precise characteristics. The indeterminacy of the word defining them is therefore counterbalanced by the illustration on the cover: The "thing" is a bull-looking creature with claws and human feet (Moebius 144).

Consequently, the Italian word "mostri" can be plausible given the monster-like characteristics of the "wild thing." More to the point, it is acceptable because the salient verbal–visual relationship in this "subtle and contemplative" cover (Marcus 245) is between the adjective "wild" and the picture of the "thing." There is a counterpoint in characterization as the "thing" is placidly sleeping and smiling. The effect is to stir the reader's curiosity. Similarly, the word "mostri" is in contrast with the visually good-natured being.

Likewise, the "topographical title" indicates an imaginary world (Nikolajeva and Scott 244) since the adverb "where" again is rather indefinite but the picture represents an exotic-looking far-away place. Hence, Italian "paese" ("land") seems to be an effective choice in the two translations because it evokes Alice's Wonderland or Pinocchio's Toyland.

Even if the two Italian translations share the same title, they do differ in terms of type font. Sendak used an Ed Interlock font, a typically American font type found mostly in television and advertisement (Boyer 41). This homage to the American popular culture that was part of Sendak's childhood was modified in the different French editions. In the first Italian translation by Emme Edizioni, the font of the second French edition by L'école des loisirs, a more classical and romantic type, was kept, but it was colored in black. As observed by Castagnoli, the second Italian translation, mirroring the publisher's philological translational project, issued the book with Sendak's original font and book jacket, counterbalancing in this way the decision to use the same title of the first Italian translation that is not philologically adherent to the source text. Different fonts are also used within the text: a larger one in Porta's translation, the same as the source text in Adelphi's new translation.

The name of the translator of the first translation appears under the title in the title page. The necessity to give salience to the name of the Italian translator probably depends on him being a poet. Somewhat differently, in the second translation, the name of the translator is written together with all the small copyright information and appears in the title page with the name of the publishing house. Again, consistently with the aims of retranslation, relevance is given to the text.

The Voice of the Translator and of the Adult Reading Aloud

The verbal and visual structure of *Where the Wild Things Are* is very sophisticated. Grounded on a previous analysis carried out on the Swedish, German, and Finnish translations of the Sendakian *chef d'œuvre* (Oittinen, "Where the Wild Things Are"), Oittinen et al. widen the analysis to include the Spanish translation and the first Italian translation. While recognizing the importance of the interplay between texts and images, their investigation principally focuses on the sentence structure and length and the deriving rhythm. Oittinen explains,

> [T]he rhythm created through alternating the sentence length is very important and should be conveyed in the translations of the book as well, as the sentence length is an important part of the narration. Sentence length is also a visual factor influencing the aloud-reader's task as well as the contents and the style of the story. ("Where" 135)

However, the rhythm of picturebooks and their performance on the part of the adult aloud-reader is construed concurrently by the verbal texts, the images, and other graphic elements (Oittinen, *Translating*; O'Sullivan, *Comparative*). The following analysis will try to account for the translational choices of the two Italian versions. All the factors determining the reading aloud situation will be taken into account.

The first long sentence describes Max's row with his mother over three double-spreads with the written text on the left and the picture on the right:

> The night Max wore his wolf suit and made mischief of one kind/
> and another/
> his mother called him "WILD THING!"
> and Max said "I'LL EAT YOU UP!"
> so he was sent to bed without eating anything.

The first Italian translation adds a full stop dividing the sentence into two:

> Quella sera Max si mise il costume da lupo e ne combinò di tutti i colori/
> e anche peggio./
> La mamma gli gridò: 'MOSTRO SELVAGGIO!'
> e Max le rispose "E IO TI SBRANO"
> Così fu cacciato a letto senza cena.[1]

Actually, the capital letter of the Italian translation of "so" implies three sentences. From a lexical point of view, the relationship between the scolding of Max's mother and the title is preserved. Full capitals are used for Max's reply, indicating the adult aloud-reader is considered able to decipher this typographical instruction, further supported by the choice of the verb "gridò" ("to shout"). "I'll eat you up" is translated into a semantically stronger "io ti sbrano" ("I'll chew you up") so that the exclamation mark of the source text disappears in the target text. Nevertheless, while losing the emphasis on the act of "eating" that characterized the book, the verb opens up a new semantic correlation. This is strengthened by the primary meaning of the Italian verb "cacciare" ("to hunt") in the expression "fu cacciato a letto," which evokes Max's disguise as a wolf, an animal, and to the adjective "wild." Furthermore, the verb "si mise" ("put on") is used to translate "wore": this shows the translator's attention to the readability of the text.

The second Italian translation clearly marks the third sentence with a full stop:

> Quella sera Max indossò il costume da lupo e ne combinò una delle sue/ e poi un'altra./
> "SELVAGGIO!" gridò la mamma. "E ALLORA IO TI MANGIO!" urlò Max.
> Così fu spedito a letto senza cena.[2]

Interestingly, if the important association between the title and Max's mother's reproach is lost, the *fil rouge* of the verb "to eat" with the monsters' declaration of love at the end of the book is maintained. The binary semantic relationship "eat–love" is then restored.

Both translations stop the flow of the story, hence its rhythm (Oittinen et al. 151). If the second translator states that choosing the parataxis and the free verse (Topi, "Ritorno") was in line with the need to create a polished text, it is nonetheless probable that Porta considered the sentence too long for the adult aloud-reader. Indeed, within the framework of the reading aloud situation, the second translator postpones the speaking subject in the dialog between Max and his mother, diverging from the source text. Sendak's original text and the first translation adopt an oral strategy that involves the anticipation of the identity of the speaking person, referred to as a form of oralization of the written text in the pedagogy of reading (Cardarello).

With just one single sentence covering five pages, the book visually and verbally describes a "long, neverending time, which goes on and on until Max enters the land where the wild things are[3]: Max takes a boat there, and looking at the illustration, the reader can see and almost feel the sudden stop, the boat striking against the shore of the land of where the wild things are" (Oittinen, "Where" 136). Porta[4] and Lisa Topi[5] keep the same long sentence, but partially use the conjunction "and" (with Topi adding a comma), hence partly rendering the climatic construction (Oittinen, *Meta*) typical of the source text.

In this long narration, there is a dash in the original text, which signals a pause—leaving the image on the right side with the task of narrating the huge growth of the forest in the protagonist's room, representing Max's increasing rage ("That very night in Max's room a forest grew/and grew—/"). Visually, the pictures become bigger and bigger, indifferent to the white frame, as his rage grows. William Moebius argues, "as Max's universe expands from the small framed picture of himself in a room to the unframed double-spread of the place where the wild things live" (150). In contrast, Porta does not leave anything implicit in the Italian text and explicates the punctuation mark. A figure of speech of children's literature is in fact employed to convey a similar sense of duration—the epizeuxis—by repeating three times the verb "crebbe" ("e crebbe crebbe crebbe," meaning "and grew, grew, grew"). It is certainly used "to create rhyme and emphasis" (Oittinen et al. 157), favoring the reading aloud situation (Alunni), but it also subsumes a different implied adult aloud-reader, that is, one who needs help when illustrations narrate the story and needs the verbal text to accompany the visual growth of the forest.

Furthermore, the abstract description "and the walls became the world all around" turns into a more concrete description of what happens in the first translation ("le pareti si trasformarono in foresta," meaning "the walls became a forest"). "[T]he last part of the analyzed sentence is the change in the length of Max's boat ride from weeks to months (mesi) to render the idea of the long journey clearer" (Oittinen et al. 157). The reader is also supported in his/her reading through the two repetitions of the adverb "perfino" ("even"), which highlights the exceptionality of the event, the rhyme between the "mare" and "navigare" ("sea" and "sail"), and the diminutive "barchetta" ("tiny boat"), a typical feature of Italian children's literature. Consequently, the process of interpretation of the visual and the verbal is facilitated thanks to some additions. Instead, the second translator employs the suspension points that leave the image to recount the actions ("e crebbe…," meaning "and grew…"). Her translation is adherent to the source text, adapting Sendak's line in "dalle pareti entrò il mondo" ("the world entered from the walls"), except for the different relationship between the verbal and the visual when she takes the presence of the sea for granted (Alunni) as the illustration does by writing,

> e sulle creste del mare apparve una barca tutta per Max
> che navigò giorno e notte.[6]

After Max meets the wild things and becomes their king, there is the climax of the story and of Max's outburst in the three central and wordless double-spreads anticipated by the verbal text: "And now," cried Max, "let the wild rumpus starts!". The adult aloud-reader is then supposed to perform together with the child the visual instructions (Sezzi, "Bridging the Sensorial Gaps") given by the illustrations that cover three pages and are a fully animated scene. The word "rumpus" of the previous page prepares the readers. Porta's and Topi's translations of this word are the heart of the debate. Familiar Italian words (in Italian "putiferio," "finimondo," or "cagnara") are turned down in favor of "ridda," which indicates an old round dance and (at a figurative level) a whirling and unrestrained movement.[7] The word "ridda" discloses Porta's education and poetic background, thus making his voice audible, and also the fact that the first translation is outdated (Topi, "Ritorno"; Alunni). Therefore, the second translator decides for a more common term, "finimondo"—deriving from the expression "end of the world" and indicating confusion and hassle—preceded by the verb "scateniamo," which means both "to start" and "to go wild."[8] Furthermore, Topi underlines how the word she has chosen coherently conveys this idea of "an anarchic destructive impetus ... an elsewhere in which the wild things can be tamed" ("Ritorno"),[9] also effectively referring back to the world that entered in Max's room at the beginning of the picturebook.

Against the claims that the "ridda" is obsolete, (the word was almost certainly unknown in Italy in the late 1960s), it must be said that the word "rumpus," referring to "A riot, an uproar, a disturbance; a row, a noisy dispute," seems not to be very frequent too.[10] It is also associated with the expression "rumpus room," the recreation room for children. It might refer to Max's dispute with his mother and his consequent rebellion, but it also invites the child readers to freely play as Max does in the three central pages. The liberating dance therein illustrated has its correlation in the Italian term used by the first translator. Relying on Sendak's sophisticated collaboration between text and illustrations, Porta chooses an equally sophisticated term that phonologically echoes the English one and that is simultaneously ready, in the transmediating process, to be semantically charged by the illustrations that lead the adult aloud-reader to wildly dance together with the child listener/viewer. After having held the hand of the adult aloud-reader in the performance of the text, after he/she has understood the mechanism, Porta now leaves him/her free to interpret Sendak's "richesse iconotextuelle" (Nières-Chevrel, "Les albums" 77), his iconotextual richness, as Max is untutored in his wild dance. The climax of the story is literally the climax of the adult aloud-reader's performance in the transmediation process in which "we interpret the text in terms of the pictures and the pictures in terms of the texts in a potentially never-ending sequence" (Sipe 102).

Maculotti, while highlighting the historical (adult) fans' preference for Porta's lexical choice, notices that its more representational power has to be ascribed to the choice of the verb "attaccare"—whose meaning is both "to start" and also "to attack" (Fiamma)—thus relating to the adjective "wild" and its interrelated semantic network: "e lo fecero re di tutti i mostri selvaggi. / 'E adesso' urlò Max 'attacchiamo la ridda selvaggia!'" The Italian entire expression epitomizes the essence of these double-spreads, which "probably comprise the best-thumbed pages in contemporary children's literature" and where "Max and four large wild things dance and bay at the moon" (Lanes 92).

After the rumpus, Max sends "the wild things off to bed." He feels lonely and gives up being king.[11] According to Spitz, "He relinquishes the fantasy of unlimited power and total independence" (133). His shout is not so loud anymore (capital letters are not used); his rage calmed down. The two translations[12] are adherent to Sendak's texts except for some differences consistent with their previous translation choices. Porta's differences include using the verb "cacciare" when Max sends the wild things off to bed, which reasserts the main semantic field, and the omission of the reference to the "world" in "Fu allora che odorò tutt'intorno un profumo di cose buone da mangiare" ("It was then that he smelled good things to eat all around"). Topi's deviations include her use of the word "mondo" ("world") and the addition of the adverb "terribilmente" ("terribly") in "qualcuno che lo amasse terribilmente" ("someone who terribly loved him")—a loan from the French translation, as she admits (Topi, "Ritorno"), that coherently evokes the adjective "terribile" and also pays homage to Sendak's translation.

Differences are to be found in the following double-spread, where Max is leaving and the wild things do not want him to go away:

> But the wild things cried, "Oh please don't go –
> We'll eat you up – we love you so!"
> And Max said, "No!"

The meaning of the relation between "eating" and "love" is now blatant, as is the connection with the destruction of the object of love (Spitz; Terranova). Porta's initial solution of "sbranare" ("to chew up") is now substituted by the verb "mangiare" ("to eat"): The dichotomy of cannibalism is then emphasized. Furthermore, there is a marked inversion between the adverb "così tanto" ("so much") and the verb "ti amiamo" ("we love"):

> Ma i mostri selvaggi gridarono: "Oh, non andartene –
> Noi ti vogliamo mangiare – così tanto ti amiamo!"
> Max rispose: "No!"

This choice is not dependent on Porta's poetic instinct (Alunni) but on the rhyming lines of the source text ("go," "so," and "no"), difficult to render in the Italian language. It is a sort of compensation, probably like the rhyme between "mare" and "navigare."

The voice of the second translator is heard in this short dialogue. The connection between "love" and "eating" is perhaps thought not to be so clear so that the consequential relation is explicated by anticipating "Ti amiamo così tanto!" ("We love you so much!") with respect to "Ti mangeremmo!" ("We would eat you!"):

> I mostri selvaggi lo supplicavano: "Non te ne andare!
> Ti amiamo così tanto! Ti mangeremmo!"
> "No!" strillò Max.

A more precise verb is used: Instead of "cry," the translator uses "supplicare" ("to beg"). Max's trip back home is narrated in the same way as the journey to where the wild things are. Coherently, the two translations do the same. The last lines of the source text are important:

> and into the night of his very own room
> where he found his supper waiting for him/
> and it was still hot.

These lines are on a blank page. His supper is waiting for him: "His invisible mother has survived his assaults. She still loves him; she is still feeding him" (Spitz 134). The verb "was waiting" is important. This is used in the first Italian translation,[13] while it is unfortunately lost in the second one,[14] replaced by "trovò" ("found"), even though it follows the syntactical structure of the original.

Conclusions

Fifty years after the first translation of *Where the Wild Things Are* into Italian, a second translation was re-issued. The reasons behind the publication were merely dependent on the re-negotiation of Sendak's copyrights. As Koskinen and Paloposki underline, retranslation is a "polemical act by nature" (27). The debate on the second translation took place especially on the web and social media, just after Adelphi publishing house announced their new translation with the name of the re-translator on twitter in January, after it went out of stock in 2016. The expectations were enormous, especially with regard to the translation of the word "ridda" (rumpus), in Anna Castagnoli's Facebook page, "Le figure dei libri," dedicated to picturebooks for a specialist and non-specialist audience, one of the comments was "I'm a bit concerned about this new translation. I'm attached to the classic text. … I wonder if the mythical wild 'ridda' will be kept." Again, after the post on the publication, "Translation is a difficult task …Lisa Topi is really faithful to the original text and to its simplicity … my heart bumped when I read 'ridda selvaggia,' this for emotional attachment I cannot but support Porta's translation."[15] Thus, contextual voices of the adult aloud-readers, teachers, and parents, also took part in the debate.

While the motives behind the retranslation were attributed to the dynamics of the publishing market, many of the contextual voices (including the one of the second translator) have legitimized the "new translation"—conforming to two of the reasons for retranslating identified by Enrico Monti: The first translation is considered out-dated since it employs an old-fashioned language; it is also a creative translation given the translator's status of poet, writer, and scholar. One website states that Lisa Topi's translation is "beautiful and studied" and that Porta's historical translation is "poeticized" (Terranova). Another affirmed that Porta's translation has the merit of giving the readers a text "happily influenced by the poetic inclination of the translator," while Topi's merit is "to bring the readers closer to the clarity of Sendak's language" (Alunni).

Hence, the source-oriented approach of the second translator is emphasized. It is also aligned with Adelphi, a refined publishing house for literature addressed to adults that launched an equally refined series of children's literature. Nevertheless, as the analysis of the different voices shows, the translator adopted some solutions that are not so philologically oriented: She tries to convey Sendak' style, creating her own semantic references and her own rhythm, paying attention to the adult reading aloud, hence generating varied transmediation processes. She was "brave as a lion," as one of the comments in the Facebook page "Le figure dei libri" highlights, because she had to face a canonical text with a very famous and successful translation, even within the limits imposed by the copyrights on the first translation.

However, within the choir of the contextual voices, the textual ones in Porta's translation show that his translation is not "poetic" and detached from the source text. Sendak's text itself has assonances, rhymes, and a specific structure. Published in the late 1960s, when the modern picturebook was just introduced in Italy by the Emme Edizioni, Porta takes care of the adult reading the story aloud, who probably was not too familiar with the genre at that time. Nonetheless, Porta is simultaneously loyal to Sendak's poetics—where the verbal, the visual, and performance entwine—and aware of the transmediation process subsumed:

> In Sendak's picturebooks, as in all picturebooks, there are some difficulties linked to the oralisation of the translated text and to the interactions between the text and the pictures .... Like it always happens with Sendak's picturebooks, even the simplest picturebooks reveal to be of a misleading simplicity.[16] (Nières-Chevrel, "Les albums" 69)

In particular, as demonstrated by the analysis, Porta's translation is neither "poetic" nor "not literal," but it is "performance-oriented," adopting Oittinen's expression. And its recognized "force" (Fiamma) probably lies in this. Nevertheless, Italian children will adhere to the new translation pact, fully enjoying the retranslation.

Notes

1. Back-translation: "That night Max wore the wolf suit and made mischiefs of all colours/and even worse./The mom shouted at him: 'WILD MONSTER!' and Max answered: "And I chew you up" So he was sent to bed without dinner" (Oittinen et al. 2017, p. 152).
2. Back-translation: "That night Max worse the wolf suit and made one mischief of his kind/and another/ 'WILD!', his mother shouted. 'AND

THEN I'LL EAT YOU UP!', Max shouted. So he was sent to bed without dinner."

3. "That very night in Max's room a forest grew/
and grew—/
and grew until his ceiling hung with vines
and the walls became the world all around/
and an ocean tumbled by with a private boat for Max
and he sailed off through the night and day/
and in and out of weeks
and almost over a year
to where the wild things are."

4. "Nella camera di Max quella sera una foresta crebbe/
e crebbe crebbe crebbe/
crebbe fino al soffitto ormai fatto di rami e foglie
e pure le pareti si trasformarono in foresta/
e si formò perfino un mare
con sopra una barchetta tutta per Max
che giorno e notte si mise a navigare
e navigò in lungo e in largo
per mesi e mesi
infine dopo un anno o poco più
giunse nel paese dei mostri selvaggi

Back-translation: "In Max's room that night woods grew/ and grew grew grew/ grew till the ceiling made by branches and leaves by now and even the walls became a forest/ and even a sea appeared with above a tiny boat entirely for Max who day and night started sailing/ and he sailed far and wide for months and months eventually after a year or a little more he reached the land of the wild monsters."

5. "Quella notte nella camera di Max spuntò una foresta/
e crebbe.../
crebbe finché il soffitto si coprì di rami
e dalle pareti entrò il mondo/
e sulle creste del mare apparve e sulle creste del mare apparve una barca tutta per Max
che navigò giorno e notte/
navigò intere settimane,
per un anno e poco più
fino al paese dei mostri selvaggi."

Back-translation: "That night in Max's room a forest appeared/and grew.../grew untill the ceiling was covered by branches and the world entered from the walls/and on the crests of the sea a boat entirely for Max appeared who sailed day and night/he sailed for entire weeks, for a year and a little longer until the land of the wild monsters."

6. Back-translation: "and on the crest of the sea a boat entirely for Max appeared that sailed day and night."
7. "e lo fecero re dei mostri selvaggi./ 'E adesso' urlò Max 'attacchiamo la ridda selvaggia!'"
8. "Lo fecero re dei mostri./ 'E ora,' gridò Max 'scateniamo il finimondo!'"
9. My translation.
10. "This word belongs in Frequency Band 4. Band 4 contains words which occur between 0.1 and 1.0 times per million words in typical modern English usage. Such words are marked by much greater specificity and a wider range of register, regionality, and subject domain than those found in bands 8-5. However, most words remain recognizable to English-speakers, and are likely be used unproblematically in fiction or journalism. Examples include overhang, life support, register, rewrite, nutshell, candlestick, rodeo, embouchure, insectivore" (http://www.oed.com/view/Entry/168864?rskey=pAHnEW&result=1&isAdvanced=false#eid).
11. "'Now stop!' Max said and sent the wild things off to bed
    without their supper. And Max the king of all wild things was lonely
    and wanted to be where someone loved him best of all./
    Then all around from far away the cross the world
    he smelled good things to eat
    so he gave up being king of where the wild things are."
12. Porta: "'Ora basta!' disse Max e cacciò i mostri selvaggi a letto senza
    cena.
    E Max, il re di tutti i mostri selvaggi si sentì solo e desiderò
    di essere in un posto dove c'era qualcuno che lo amava
    più di ogni altra cosa al mondo./
    Fu allora che odorò tutt'intorno
    un profumo di cose buone da mangiare"

    Back-translation: "'Now stop!' Max said and he hunted the wild monsters off to bed without supper. And Max, the king of all wild monsters, felt lonely and wished to be with someone who loved him ore than any one else in the world./ It was then that he smelled all arounda perfume of good things to eat that came from far away so he gave up being the king of the land of the wild monster."

    Topi: "'Ora basta!' gridò Max mandando i mostri a letto senza cena.
    E Max il re dei mostri selvaggi si sentì solo,
    avrebbe voluto essere con qualcuno che lo amasse terribilmente./
    Poi, all'improvviso, dall'altra parte del mondo sentì arrivare
    un buon profumo di cose da mangiare
    e decise che non voleva più essere il re dei mostri selvaggi."

Back-translation: "'Now stop!' cried Max sending the monsters off to bed without their supper. And Max the king of the wild monsters felt lonely, he wanted to be with someone who terribly loved him./Then, suddenly, from the other part of the world he smelled a good perfume of things to eat and decided that he didn't want to be the king of the wild monsters."

13. "Finché tornò a quella sera nella sua stanzetta
dove trovò la cena ad aspettarlo/
che era ancora calda"
14. "fino a quella notte dove in camera sua
trovò la cena/
ancora calda"
15. My translations.
16. My translation of "Il y a dans le albums de Sendak, comme dans tous les albums, des difficultés de traduction, liées a l'oralisation du texte traduit et aux interactions du texte e de l'image [...]. Comme toujours chez Sendak, même les albums le plus simples se révèlent d'une trompeuse simplicité" (Nières-Chevrel, "Les albums de Maurice Sendak" 69).

Works Cited

Alunni, Lorenzo. "Nel paese dei mostri selvaggi: la nuova traduzione." *il lavoro culturale*, 2018, http://www.lavoroculturale.org/paese-mostri-selvaggi/. Accessed 18 September 2018.

Alvstad, Cecilia. "The Translation Pact." *Language and Literature*, vol. 23, 2014, pp. 270–284.

Alvstad, Ceclilia and Alexandra Assis Rosa. "Voice in Retranslation. An Overview and Some Trends." *Target*, vol. 27, no, 1, 2015, pp. 3–24.

Berman, Antoine. "La retraduction comme espace de la traduction." *Palimpsestes*, vol. 1, edited by Bensimon Paul and Coupaye Didier, Presses de la Sourbonne Nouvelle, 1990, pp. 1–7.

Boyer, Löic. "Contes de Grimm et culture pop! Ou Maurice Sendak montre toujours autre chose." *Max e les maximonstres à 50 ans. Réception et influence des œuvres de Maurice Sendak en France et en Europe*. BNF Centre nationale de la littérature pour la jeunesse, 2014, pp. 6–15.

Cabaret, Florence. "Introduction." *La Retraduction en littérature de jeunesse. Retranslating Children's Literature*, edited by Virginie Douglas and Florence Cabaret. Oxford: Peter Lang, 2014, pp. 11–19.

Cardarello, Roberta. *Libri e bambini. La prima formazione del lettore*. Firenze: La Nuova Italia, 1995.

Castagnoli, Anna. "Da Mickey Mouse ai mostri selvaggi: genesi di un capolavoro." *Le figure dei libri*, 29 March 2018, http://www.lefiguredeilibri.com/2018/03/29/maurice_sendak_adelphi/. Accessed on 18 September 2018.

Chatman, Seymour. *Storia e discorso. La struttura narrativa nel romanzo e nel film*. 1978. Translated by E. Graziosi. Piacenza: Pratiche, 1998.

Christensen, Nina. "Mature Children and Infantile Adults: The Depiction of Child-Parent relationships in Danish Picture Books." Conference on Reading Pictures: Art, Narrative & Childhood, Cambridge, 2000, http://www.cfb.dk/site.aspx?p=69.

Desarthe, Agnès. "Traduire *Nutshell Library*: contraintes et Libertés." *Max e les maximonstres à 50 ans. Réception et influence des œuvres de Maurice Sendak en France et en Europe*. BNF Centre nationale de la littérature pour la jeunesse, 2014, pp. 78–84.

Dollerup, Cay. "Translating for Reading Aloud." *Meta*, vol. 48, May 2003, pp. 81–103.

Doonan, Jane. *Looking at Pictures in Picture Books*. Stroud, Gloucestershire: Thimble Press, 1993.

Douglas, Florence. "Conclusion." *La Retraduction en littérature de jeunesse. Retranslating Children's Literature*, edited by Virginie Douglas and Florence Cabaret. Oxford: Peter Lang, 2014, pp. 317–329.

Douglas, Virginie, and Florence Cabaret, editors. *La Retraduction en littérature de jeunesse. Retranslating Children's Literature*. Oxford: Peter Lang, 2014.

Fiamma, Andrea. "'Nel paese dei mostri selvaggi.' Torna il capolavoro di Maurice Sendak." *Fumetto Logica*, 22 February 2018, http://www.fumettologica.it/2018/02/nel-paese-dei-mostri-selvaggi-maurice-sendak/. Accessed on 18 September 2018.

Fochesato, Walter. *Libri illustrati: come sceglierli?* Milano: Mondadori, 2000.

Hermans, Theo. "The Translator's Voice in Translated Narrative." *Target*, vol. 8, no. 1, 1996, pp. 23–48.

Klingberg, Göte. *Children's Fiction in the Hands of Translators*. Studia psychologica et pedagogica, Series Altera. Lund: CWK Gleerup, 1986.

Koskinen, Kaisa and Outi Paloposki. "Anxieties of Influence. The Voice of the First Translator in Retranslation." *Target*, vol. 27, no. 1, 2015, pp. 25–39.

Kress, Gunter and Theo van Leeuwen. *Reading Images: The Grammar of Visual Design*. Routledge, 1996.

Lanes, Selma G. *The Art of Maurice Sendak*. Abradale Press, 1980.

Lathey, Gillian. "Introduction." *The Translation of Children's Literature: A Reader*, edited by Gillian Lathey. Clevedon: Multilingual Matters, 2006, pp. 1–12.

Lewis, David. "Going Along with Mr Gumpy: Polysystemy & Play in the Modern Picture Book." *Signal*, vol. 80, 1996, pp. 105–119.

———. *Reading Contemporary Picturebooks: Picturing Text*. Routledge Falmer, 2001.

Maculotti, Matteo. "Il ritorno dei mostri selvaggi di Maurice Sendak." https://www.bambinietopi.it/2018/04/mostri-selvaggi-maurice-sendak.html. Accessed on the 18 September 2018.

Marcus, Leonard S. "Let the Wild Rumpus Start: Maurice Sendak as Artist, Storyteller and Psychologist." *Max e les maximonstres à 50 ans. Réception et influence des œuvres de Maurice Sendak en France et en Europe.* BNF Centre nationale de la littérature pour la jeunesse, 2014, pp. 6–15.

Moebius, William. "Introduction to Picturebook Codes." *Word & Image: A Journal of Verbal/Visual Enquiry*, vol. 2, no. 2, April–June 1986, pp. 141–158.

Monti, Enrico. "La Retraduction, un état des lieux." *Autour de la retraduction. Perspectives littéraires européennes*, edited by Enrico Monti and Peter Schnyder. Paris: Orizons, coll. "Universités", 2012, pp. 9–25.

Nières-Chevrel, Isabelle. "Traduire in the Night Kitchen, ou de la difficile lecture d'un album." *Meta*, vol. 48, no. 1–2, 2003, pp. 154–164.

———. "Les albums de Maurice Sendak: quels problèmes de traduction." *Max e les maximonstres à 50 ans. Réception et influence des œuvres de Maurice Sendak en France et en Europe.* BNF Centre nationale de la littérature pour la jeunesse, 2014, pp. 69–77.

Nikolajeva, Maria. "The Verbal and the Visual: The Picturebook as a Medium." *Children's Literature as Communication*, edited by Roger D. Sell. Amsterdam: John Benjamins Publishing Company, 2002, pp. 85–107.

Nikolajeva, Maria and Carole Scott, editors. *How Picturebooks Work.* Routledge, 2006.

Nodelman, Perry. *Words About Pictures: The Narrative Art of Children's Picture Books.* Athens: University of Georgia Press, 1988.

Oittinen, Riitta. "The Dialogic Relation Between Text and Illustration: A Translatological View." *TexTconText*, vol. 5, 1990, pp. 40–53.

———. "The Verbal and the Visual: On the Carnivalism and Dialogics of Translating for Children." *Compar(a)ison*, vol. 2, 1995, pp. 49–65.

———. *Translating for Children.* New York: Garland Publishing, 2000.

———. "Where the Wild Things Are: Translating Pictures Books." *Meta*, vol. 48, May 2003, pp. 128–141.

———. "Change and Renewal: Translating the Visual in Picture Books." *Change and Renewal in Children's Literature*, edited by Theo Van der Walt. Westport: Praeger Publisher, 2004, pp. 171–181.

Oittinen, Riitta, Anne Ketola, and Melissa Garavini. *Translating Picturebooks: Revoicing the Verbal, the Visual and the Aural for a Child Audience.* Routledge, 2017.

O'Sullivan, Emer. "Translating Pictures." *Signal*, vol. 90, 1998, pp. 167–175.

———. "'Anything to Me Is Sweeter…' British Translations of Heinrich Hoffman's Struwwelpeter." *Princeton University Library Chronicle*, vol. 62, no. 1, 2000, pp. 59–71.

———. *Comparative Children's Literature*. Translated by A. Bell. Routledge, 2005.

Pallottino, Paola. *Storia dell'illustrazione italiana. Libri e periodici a figure dal XV al XX secolo.* [A History of Italian Illustrations. Illustrated Books and Periodicals from 15th to 20th Century]. Bologna: Zanichelli, 1988.

Pederzoli, Roberta. "Les retraductions des contes de Perrault en Italie au XXIe siècle. Entre quête de legitimation et adaptation au(x) destinataire(s)." *La Retraduction en littérature de jeunesse. Retranslating Children's Literature*, edited by Virginie Douglas and Florence Cabaret. Oxford: Peter Lang, 2014, pp. 53–68.

Schiavi, Giuliana. "There Is Always a Teller in a Tale." *Target*, vol. 8, no. 1, 1996, pp. 1–21.

Schwarcz, Joseph H. *Ways of the Illustrator: Visual Communication in Children's Literature*. Chicago: American Library Association, 1982.

Schwarcz, Joseph H. and Chava Schwarcz. *The Picture Book Comes of Age*. American Library Association, 1984.

Sezzi, Annalisa. "Bridging the Sensorial Gaps: Theory and Practice in Translating the Voice of the Adult Aloud Reader in Pre-school Picturebooks." *Thinking Translation: Perspectives from Within and Without*. Conference Proceedings Third UEA Postgraduate Translation Symposium, R. Hyde Parker and García K. Guadarrama (a cura di), Brown Walker Press, 2008, pp. 157–174.

———. "Mediating Between Childhood and Adulthood: The Translation of Picture Books." *La geografia della mediazione linguistico-culturale* [The Geography of Linguistic-Cultural Mediation], edited by Donna R. Miller and Ana Pano. Atti del convegno CeSLiC, Quaderni del Ceslic, 2009, pp. 192–209, http://amsacta.cib.unibo.it/2626/.

———. "Bariery literatury dla dzieci. Recepcja ksiazek obrazkowych we wloszech a kwestia glosnej lektury." *Przekładaniec*. Półrocznik Katedry UNESCO do Badań nad Przekładem i Komunikacja Miedzykulturowa UJ, 2-1 (22–23), 2010, pp. 226–244.

Shavit, Zohar. *Poetics of Children's Literature*. The University of Georgia Press, 1986.

Shulevitz, Uri. *Writing with Pictures. How to Write and Illustrate Children's Books*. New York: Watson-Gullip Publications, 1985.

Sipe, Lawrence R. "How Picture Book Work: A Semiotically Framed Theory of Text-Picture Relationships." *Children's Literature in Education*, vol. 29, no. 2, 1998, pp. 97–108.

Spitz, Ellen H. *Inside Picture Books*. Yale University Press, 1999.

Suhor, Charles. "Towards a Semiotics-Based Curriculum." *Journal of Curriculum Studies*, vol. 16, 1984, pp. 247–257.

Tabbert, Reinbert. "The Impact of Children's Books—Cases and Concepts." *Responses to Children's Literature*, edited by Geoff Fox. Sauer, 1980, pp. 34–58.

Terranova, Nadia. "Maurice Sendak il ritorno dei mostri." *gli asini*, 2018, http://gliasinirivista.org/2018/04/maurice-sendak-ritorno-dei-mostri/. Accessed on 18 September 2018.

Terrusi, Marcella. *Albi illustrati. Leggerre, guardare, nominare il mondo nei libri per l'infanzia*. Roma: Carrocci editore, 2012.

Topi, Lisa. "Ritorno al paese dei mostri selvaggi." *Tradurre*, 2018, https://rivistatradurre.it/2018/05/ritorno-al-paese-dei-mostri-selvaggi/. Accessed on 18 September 2018.

———. "Nel paese dei mostri selvaggi Adelphi ripropone il capolavoro di Sendak." Interview by Andrea Fiamma, 2018, https://www.andmagazine.it/index.php/component/content/article/13-2018/rubrica-di-cultura/28-nel-paese-dei-mostri-selvaggi-adelphi-ripropone-il-capolavoro-di-sendak. Accessed on 18 September 2018.

Topipittori. "Signor Calasso, Where Are the Wild Things?" *Topipittori*, 2017, https://www.topipittori.it/it/topipittori/signor-calasso-where-are-wild-things. Accessed on 18 September 2018.

Treccani vocabolario online: http://www.treccani.it/.

Valentino Merletti, Rita. "Quando si cominciò a raccontare con parole e immagini." *Liber: Libri per bambini e ragazzi*, vol. 61, Campi Bisenzio, 2004, pp. 22–23.

van Coillie, Jan. "How Hard Is to Play the Translator's Game." *Marvels and Tales*, vol. 2, 2014, pp. 346–365.

———. "Nibble, Nibble Like a Mouse/Who Is Nibbling at the Source Text's House. Retranslating Fairy Tales: Untangling the Web of Causation." *La Retraduction en littérature de jeunesse. Retranslating Children's Literature*, edited by Virginie Douglas and Florence Cabaret. Oxford: Peter Lang, 2014, pp. 39–52.

van Coillie, Jan and Walter P. Verschueren, editors. *Children's Literature in Translation: Challenges and Strategies*. Manchester: St Jerome, 2006.

van Lierop-Debrauwer, Helma. "Hybridity in Picturebooks." *The Routledge Companion to Picturebooks*, edited by Bettina Kümmerling-Meibauer. Routledge, 2018, pp. 81–90.

Winters, Kari-Lynn, et al. "Performing Picture Books as Co-authorship." *More Words About Pictures Current Research on Picture Books and Visual/Verbal Texts for Young People*, edited by Naomi Hamer, Perry Nodelman, and Mavis Reimer. Routledge, 2017, pp. 100–115.

Translating Ambiguity: The Translation of Dual Address in Children's Fantasy During the 1950s and 1960s

Agnes Blümer

Two Key Concepts in Crossover Literature: Address and Ambiguity

Most translations of children's literature are in some way concerned with their audience, the children, and their (presupposed) needs. Children's literature with crossover appeal—that is, literature that caters to both adults and children or that may even be read in different ways depending on whether a naïve or an experienced reader is reading it—presents an especially interesting case for translation studies. Texts that offer different readings have received extensive attention from children's literature scholars, who have generated a number of terms and concepts useful for analysis. In this chapter, I propose to add two new concepts that are related to points already established by children's literature and translation studies scholars: "visual context adaptation" and "generic affiliation." Looking at the adaptation mechanisms that crossover children's literature as well as its visual elements can be subject to during the translation process will shed some light both on the concept of dual address

A. Blümer (✉)
ALEKI, University of Cologne, Cologne, Germany

A. Kérchy and B. Sundmark (eds.), *Translating and Transmediating Children's Literature*, Critical Approaches to Children's Literature,
https://doi.org/10.1007/978-3-030-52527-9_15

and on the translation practices in children's literature during a certain period of time.

The most influential scholars and ideas when it comes to dual address include Zohar Shavit on the ambivalence of children's literature; Barbara Wall on the difference between single, double, and dual address; Hans-Heino Ewers, who developed the concept of "Doppelsinnigkeit" (children's literature with double meaning); Emer O'Sullivan, who wrote on "Mehrfachadressiertheit" (multiple address); and the work of Sandra Beckett and Rachel Falconer on crossover literature. O'Sullivan has also considered the effect of translation on dual audiences, most prominently through her work on the influence of the translator's voice on the implied audience. I'll be using the term "dual address" in the sense of Wall and Ewers to denote literature with two implied readers, a child (or naïve) and an adult (or experienced) reader.

Texts and Issues

My study deals with the translation of American, English, and French children's fantasy with dual address into German during the decades after the Second World War. It examines how the process of translation in this period involved changing the implied addressees or the audience to whom the work was directed. Furthermore, it considers how and why this change occurred and what patterns might be detectable in these changes.

Although the concepts of genres and literary periods may seem outdated or normative in many contexts (see Andrea Weinmann, in *Kinderliteraturgeschichten*, on the history of German children's literature and its historiography), working within these frameworks bore in this case fruitful results. Focusing on a single genre (children's fantasy), a single literary period (the post-war era), and a special case of children's literature (dual-address or crossover literature) drew a small but diverse corpus together by bringing some recurring adaptation patterns to light. Indeed, some of my findings suggest that the patterns are not confined to the six texts I examined in this project but rather reflect more general tendencies and strategies in translation, at least those done in West Germany during the 1950s and 1960s.

The two-decade period from 1950 to 1970 was a quite cohesive period for (West) German children's literature. Its beginning was marked by the reestablishment of the German book market after the disruption caused by National Socialism and the war, and it continued until the ideas surrounding the year 1968 transformed children's literature again. During this period, children's fantasy became one of the most prolific genres in Germany (Weinmann, "Geschichte der Kinderliteratur" 25), partly due to the high number of translations published.

I applied the following criteria to demarcate my corpus:

1. Only books of fantasy originally published in English or French from 1950 to 1970 were to be considered.
2. Translation of these works into German and their publication in West Germany had to occur in this period as well. In other words, the translations had to be contemporary with the source texts (as opposed to, for example, retranslations of older classics).
3. The contemporary reception of these texts had to recognize them as dual address texts, as evidenced for example in reviews (even if that specific term was not used).
4. The corpus was to show some diversity, so that it included texts for different age groups, texts with different degrees of illustration, and several different types of fantasy (such as time or miniature-world fantasy).

Following these criteria, I selected Mary Norton's *The Borrowers* (first published in 1952), Maurice Druon's *Tistou les pouces verts* (1957), Philippa Pearce's *Tom's Midnight Garden* (1958), Madeleine L'Engle's *A Wrinkle in Time* (1962), Maurice Sendak's *Where the Wild Things Are* (1963), and Eugène Ionesco's *Conte numéro 1* (1967). The corpus consists of three illustrated texts for children (*Borrowers*, *Tistou*, and *Tom*), two texts that may be characterized as picture books (*Wild Things* and *Conte*), and one text aimed at older children or teens that contains no illustrations (*Wrinkle*).

In terms of content, one text veers toward science fiction (*Wrinkle*), one is a time fantasy (*Tom*), one is an example of the miniature-world text (*Borrowers*), and one has elements of legend (*Tistou*). *Wild Things* is a picturebook fantasy, while *Conte numéro 1* is a picturebook that verges on the absurd or surreal.

The Borrowers was translated by Theresia Mutzenbecher and was published in German in 1955 as *Die Borgmännchen* (which in turn could be retranslated as "The Little Men Who Borrow"). *Tistou les pouces verts* became *Tistou mit dem grünen Daumen* ("Tistou with the

Green Thumb") when it was translated by Hans-Georg Lenzen in 1959. *Tom's Midnight Garden* was translated by Wolfram Buddecke in 1961 as *Als die Uhr dreizehn schlug* ("When the Bell Struck Thirteen"). The (first) German version of *A Wrinkle in Time* is called *Spiralnebel 101* ("Spiral Nebula 101") and was done by Martha Johanna Hofman in 1968. *Where the Wild Things Are* was translated as *Wo die wilden Kerle wohnen* ("Where the Wild Guys Live") by Claudia Schmölders in 1967. *Conte numéro 1* was translated by Herbert Asmodi in 1969 as *Geschichte Nummer 1* ("Story Number One").

In my study, I applied close readings and compared the source to the target texts—looking at both the verbal and the visual elements. I then compared the cultural contexts, considering issues ranging from the discourses on children's literature that were ongoing in each culture to the genres in existence in each culture, the respective addressees, and the translation.

Children's Fantasy and Its Translation

The nature of the fantasy genre is such that it lends itself well to dual address and hidden meanings. Its inherent polyvalence becomes multiplied when the cultural context is taken into account in its interpretation. From a cross-cultural perspective, fantasies produced during the post-war period present a quite interesting case study, for the perception of the genre differed greatly among the cultures in question here. These different genre expectations were found to influence the translation process and often lead to changes in genre, to generic affiliation.

Neither so-called high fantasy nor domestic fantasy was popular in German-language children's literature at the beginning of the period. Despite E. T. A. Hoffmann's works and the strong German fairy tale tradition, these genres seem not to have caught on early in the form of children's literature. On the other hand, in English as well as French literature (and other languages), fantasy quickly became a richly represented genre. Celebrated literary traditions were constructed, for example, by works like *The Water-Babies* (1863), *Alice's Adventures in Wonderland* (1865), *Peter Pan* (1911), *Mary Poppins* (1935), *Le Pays des trentesix mille volontés* (1928), *Patapoufs et Filifers* (1930), or *Les contes du chat perché* (1939). Not so in West Germany, where fantasy had to be imported or reimported and was met with some resistance.

For the most part, this resistance was due to the way in which children's literature was perceived in West Germany. Most influential at the time was the *Theorie des guten Jugendbuchs* movement, the "theory of the good children's book" (see Müller for a historical study on this movement). Critics, researchers, and teachers in Germany demanded that children's literature be both *good* (i.e., of literary quality) and *age-appropriate*. Of these two demands, age-appropriateness seemed to trump literary quality. As Ruth Koch wrote in 1959, "Die kindgemäße Darstellung ist jedoch mit die erste Forderung, die wir heute an ein Kinderbuch zu stellen haben" ("Age-appropriateness, however, is the first requirement that we have to place on the children's book"; 84).

Accordingly, quite a few children's books were ruled out or greatly criticized for not being age-appropriate. Mediators' discussions often centered on translated books that did not seem to fit into the age patterns established so firmly in Germany. Although suggested age levels for children's books of course existed on the book markets in France, the United Kingdom, and the United States, the division between children's literature and adult literature was much stricter in Germany. Quite often, the books that were viewed as transgressing these boundaries were fantasy texts—for example, the *Mary Poppins* series or the works of C. S. Lewis. Koch wrote in 1959,

> Andererseits wird, gerade in den englischen Kinderbüchern ..., das Phantasieland an einigen Stellen in einer Weise gedeutet und dargestellt, die nur von reifen Jugendlichen und Erwachsenen verstanden werden kann. ... Diese Deutung ist seinem [des Kindes] Wesen fremd, wir müssen sie deshalb für ein Kinderbuch ablehnen. (60f.)

> [On the other hand, especially in English children's books ..., the fantasy realm is interpreted and presented in a way that can only be understood by older teens or adults.... This kind of interpretation is alien to the nature [of the child], so we have to reject it for children's books.]

At the same time, translations for children's books were promoted in Germany after the Second World War in the hope that exposure to foreign literature would facilitate children's international understanding and prevent further nationalistic ideas from taking root again. Thus, despite reservations and complaints like Koch's, quite a few dual address fantasy books were translated. Perhaps this is the most striking aspect of these translations: the fact that they exist at all. Even though the texts were not necessarily in accordance with the post-war German idea of a "good" children's book, they got translated anyway, were read, and were sometimes shortlisted for book prizes or even won awards.

Verbal Shifts: Generic Affiliation

Still, there were concessions made to the German standard of the "good" children's book, and translations featured various adjustments to the audience regarding genre and visual presentation of the text. Many of these shifts[1] had the effect of altering the implied addressee or changing the appeal for different age groups. Interestingly, only a few of these shifts occurred at the level of language, while many are found at the typographical or pictorial level. Verbal shifts often touch on genre: in one case of adaptation, for instance, the text (*The Borrowers*) was assimilated to the fairy-tale genre so that it would be more recognizable to the German target audience. Emer O'Sullivan describes similar adaptation using the notion of "Märchenisierung" or "fairytalization" (O'Sullivan, *Kinderliterarische Komparatistik* 320). This practice instantiates what Zohar Shavit has termed systemic affiliation. As she put it, "Translation of children's literature tends to relate the text to existing models in the target system. This phenomenon … is particularly prominent in the translation of children's literature because of the system's tendency to accept only the conventional and the well known" (Shavit 115).

Indeed, in quite a few instances, the apparently novel genre of fantasy was adapted to established German genre models that seemed less ambiguous or a better fit for the suggested age levels—an adaptation practice I will name *generic affiliation*. For example, the translation of *Tom's Midnight Garden* shifted the work from its original status as a time fantasy and an intertextually rich philosophical treatment of age to become a garden adventure or detective novel. On the verbal level, this change is most visible in the chapter headings. The English chapter headings foreground the most important themes (like "By Moonlight," "By Daylight," "Through a Door," "River to the Sea," and "Time and Time Again") or radiate nostalgia ("The Forgotten Promise") and point to philosophical and religious concepts ("The Pursuit of Knowledge" and "The Angel Speaks"). In contrast, the German titles read like a table of contents belonging to a detective or adventure story—for example, "Die Geisterversammlung" ("The Gathering of Ghosts"), "Bei Tag sieht alles anders aus" ("Everything Looks Different in Daylight"), "Tom steht vor einem Rätsel" ("Tom Faces a Mystery"), "Mit Pfeil und Bogen" ("With Bow and Arrow"), "Auf den Spuren Rip van Winkels" ("Following the Trail of Rip van Winkle"), "Eine aufregende Entdeckung" ("An Exciting Discovery"), "Königin Viktoria und die ersten Hosen" ("Queen Victoria and the First Trousers"), and "Das Geheimnis der Großvateruhr" ("The Mystery of the Grandfather Clock").

This type of genre shift happens in many of the texts examined, albeit in different ways or, rather, genre directions. For example, while the German translation of *The Borrowers* features elements that resemble a fairy tale, *A Wrinkle in Time* becomes less fantasy and more science fiction, and *Where the Wild Things Are* is billed as a "Kinderbilderbuch," a picture book that is explicitly addressed to children; moreover, *Tistou* is modeled not so much on an established genre but on a very famous text that was already well known in Germany—Saint-Exupéry's *Le Petit Prince* (*The Little Prince*, translated into German in 1950 as *Der Kleine Prinz*).

My examination of the texts suggests that a third model for the types of shifts found in these translations may be proposed. In addition to Shavit's idea of a general systemic affiliation and O'Sullivan's theory of a narrower "fairytalization," I propose something in the middle: We continuously find a translation strategy that seeks to adapt texts to different genre models better known in the target culture, depending on which elements feature prominently in the source text. Let us call this strategy *generic affiliation*, the adaptation of a translation to genre patterns already established and well received in the target culture.

Generic affiliation occurred not only on the verbal level but also on the visual. The English book cover of *Tom's Midnight Garden*, featuring a dreamy nature and garden theme, transforms in the German version to a harsher image: shadows, contrasts, clocks, and an owl—suggesting adventure, suspense, and perhaps even horror. The visual shift conformed to the verbal change in genre (Fig. 1).

Visual Shifts: Visual Context Adaptation

A work's addressee is constituted not only by the words in the text itself but also by the layout, the typography, and of course the images. There are several examples of striking visual shifts in the translation corpus: Sometimes the original illustrations were replaced with new ones, as in

Fig. 1 Left: Philippa Pearce's *Tom's Midnight Garden*, Oxford University Press, 1958 (illustration by Susan Einzig). Right: the German translation, *Als die Uhr dreizehn schlug*, Westermann Verlag, Braunschweig, 1961 (illustration by Hanns and Maria Mannhart)

the translations of *Tom's Midnight Garden* and *The Borrowers*. In *Tom's Midnight Garden*, the German edition does not feature Susan Einzig's iconic pictures but provides new images by Klaus und Maria Mannhart. The new pictures touched less on the main (and adult-oriented) topics of these books—time, power, and the gaze—and leaned more to the child appropriate and the cute. Other times, when the original illustrations were retained, several of the originals were either omitted or altered. This occurred with *Tistou*, where the book's ending is rendered less ambiguous in the German version because the original last illustration by Jacqueline Duhême is suppressed and a less ambiguous image is moved to the last page.

Even punctuation and other uses of typography were tampered with. In the translation of *The Borrowers*, the single quotation marks, which were used as one of the expressive features of the prose (to denote the power relations implied for the Borrowers when being 'seen' by adults), were instead represented conventionally—that is, left out altogether. In *Where the Wild Things Are*, several words were printed in capitals to show Max's emotional state, but this typographic marker did not find its way into the German version.

These types of shifts we may term *visual context adaptations*, parallel to the "cultural context adaptations" theorized by Göte Klingberg (1986). Like the generic adaptations, they influence the way in which the texts are perceived by adults and children, and just as with the verbal shifts, most of the visual shifts tended to make the texts less ambiguous and more conventional. This is not to say that the texts completely lost their dual address or ambiguity in the translations; it is just that in most cases the ambiguity became a little less obvious.

The Development of the Translation of Dual Address in Children's Fantasy

Over the course of the studied twenty-year period, the translation practice itself developed so that children's books showed more age ambiguity by the end of the 1960s. For example, the German translation of Ionesco's picturebook retained all the pictures as well as the layout in its entirety and made very few verbal shifts, although this text, with all its absurd musings on language, may be regarded as the most ambiguous (age-wise) of all six texts chosen here.

The suppression of the all-age appeal of books written at the beginning of the post-war period in their contemporary German translations was probably done to make the texts appropriate for a German market preoccupied at that time with age levels. However, as time progressed, the book market moved toward opening age levels, and this development continues to the present day. From the 1990s onwards, when *all age* or *crossover* became a marketable quality in children's literature, many of the texts examined here were retranslated or transformed into graphic-novel or movie adaptations. These new versions, in accordance with our contemporary concept of children's literature, tend to do the opposite of the first translations: They showcase the dual appeal of the original texts. We can see this, for example, in the latest German editions (from 1999 onwards) of *Tom's Midnight Garden*, which are translated by Klaus Fritz (also the German translator of *Harry Potter*). The German edition now preserves the texts' central concepts instead of trying to make them fit into other generic patterns. This is observable in the chapter headings: "By Moonlight" is now still "Bei Mondlicht," "Time and Time Again" is "Zeit und nochmals Zeit," and "The Angel Speaks" is "Der Engel spricht." We have come full circle—probably not because translators are better now but rather because the current concept of children's literature allows for dual-address fantasy texts to be translated as such.

Note

1. I am using the term "shift" according to Bakker et al.: "Translation, like every transfer operation, involves an 'invariant under transformation.' The transformation which is occasioned by the translation process can be specified in terms of changes in respect to the original, changes which are termed 'Shifts.' The two concepts of invariant and shift are therefore interdependent, such that any classification or definition of Shifts entails a definition of the invariant" (Bakker et al. 227).

Works Cited

Children's Books

Druon, Maurice. *Tistou les pouces verts*. Illustrated by Jacqueline Duhême. Paris: Del Duca, 1957.

Ionesco, Eugène. *Conte numéro 1*. Illustrated by Etienne Delessert. Paris: Quist, 1968.

L'Engle, Madeleine. *A Wrinkle in Time*. Farrar, Straus & Giroux, 1962.

Norton, Mary. *The Borrowers*. Illustrated by Diana Stanley. J. M. Dent & Sons, 1952.

Pearce, Philippa. *Tom's Midnight Garden*. Illustrated by Susan Einzig. Oxford University Press, 1958.

Sendak, Maurice. *Where the Wild Things Are*. Story and Pictures by Maurice Sendak. Harper and Row, 1963.

Children's Books (Translations)

Druon, Maurice. *Tistou mit den grünen Daumen*. Illustrated by Jacqueline Duhême. Translated by Hans Georg Lenzen. Düsseldorf: Rauch, 1959.

Ionesco, Eugène. *Geschichte Nummer 1. Geschichten für Kinder unter drei Jahren*. Illustrated by Etienne Delessert. Translated by Herbert Asmodi. Köln: Middelhauve, 1969.

L'Engle, Madeleine. *Spiralnebel 101. Eine mehr als abenteuerliche Geschichte.* Translated by Martha Johanna Hofmann. München: Claudius, 1968.

Norton, Mary. *Die Borgmännchen.* Illustrated by C. Walter Rauh. Translated by Theresia Mutzenbecher. Freiburg: Herder, 1955.

Pearce, Philippa. *Als die Uhr dreizehn schlug.* Illustrated by Hanns und Maria Mannhart. Translated by Wolfram Buddecke. Braunschweig: Westermann, 1961.

———. *Als die Uhr dreizehn schlug.* Translated by Klaus Fritz. Hamburg: Dressler, 1999.

Sendak, Maurice. *Wo die wilden Kerle wohnen. Ein Diogenes Kinderbuch.* Translated by Claudia Schmölders. Zürich: Diogenes, 1967.

Critical Works

Bakker, Matthijs et al. "Shifts of Translation." *Routledge Encyclopedia of Translation Studies*, edited by Mona Baker and Kirsten Malmkjær. Routledge, 1998, pp. 226–231.

Beckett, Sandra L. *Crossover Fiction. Global and Historical Perspectives.* Routledge, 2009.

Blümer, Agnes. *Mehrdeutigkeit übersetzen. Englische und französische Kinderliteraturklassiker der Nachkriegszeit in deutscher Übertragung.* Peter Lang, 2016.

Ewers, Hans-Heino. *Fundamental Concepts of Children's Literature Research. Literary and Sociological Approaches.* Translated from German by William J. McCann. Routledge, 2009.

Falconer, Rachel. "Cross-Reading and Crossover Books." *Children's Literature. Approaches and Territories*, edited by Janet Maybin and Nicola J. Watson, Macmillan, 2009, pp. 66–379.

Koch, Ruth. "Phantastische Erzählungen für Kinder. Untersuchungen zu ihrer Wertung und zur Charakteristik ihrer Gattung." *Studien zur Jugendliteratur*, vol. 5, 1959, pp. 55–84.

Müller, Sonja. *Kindgemäß und literarisch wertvoll. Untersuchungen zur Theorie des 'guten Jugendbuchs' – Anna Krüger, Richard Bamberger, Karl Ernst Maier.* Peter Lang, 2014.

O'Sullivan, Emer. *Kinderliterarische Komparatistik.* Heidelberg: Winter, 2000.

———. "Narratology Meets Translation Studies, or The Voice of the Translator in Children's Literature." *The Translation of Children's Literature. A Reader*, edited by Gilian Lathey. Multilingual Matters, 2006, pp. 98–109.

Shavit, Zohar. *Poetics of Children's Literature.* University of Georgia Press, 1986.

Wall, Barbara. *The Narrator's Voice. The Dilemma of Children's Fiction.* Macmillan, 1991.

Weinmann, Andrea. "Geschichte der Kinderliteratur der Bundesrepublik nach 1945." *Kinder- und Jugendliteratur der Gegenwart. Ein Handbuch*, edited by Günter Lange. Baltmannsweiler: Schneider Hohengehren, 2011, pp. 13–57.

———. *Kinderliteraturgeschichten. Kinderliteratur und Kinderliteraturgeschichtsschreibung in Deutschland seit 1945*. Peter Lang, 2013.

"Maxima Debetur Puero Reverentia": The Histories and Metamorphoses of Latin Translation in Children's Literature

Carl F. Miller

When I mentioned to colleagues that I was working on a project involving Latin translation and children's literature, most assumed it was regarding the translation of classical and medieval works into English for contemporary children and expressed surprise (even confusion) when they were told otherwise. The other frequent reaction I received while carrying Latin translations of *The Cat in the Hat* and *Winnie-the-Pooh* in public was amusement—most often stemming from the odd visual juxtaposition of the popular field of children's literature with the intellectually elite subject of Latin. Even students in my children's literature seminars who do not bat an eye at titles in Spanish, Russian, and Chinese invariably snicker when introduced to familiar children's books translated into Latin, often initially believing that I am playing a scholarly joke on them.

In fact, there is a rich history over the past 150 years of translating modern children's literature into Latin, an odd reverse dynamic of translating contemporary works into a presumably dead language. As the Latin translation of children's literature in most of the world has been comparatively limited in production, this study will focus on Western Europe

C. F. Miller (✉)
Palm Beach Atlantic University, West Palm Beach, FL, USA

A. Kérchy and B. Sundmark (eds.), *Translating and Transmediating Children's Literature*, Critical Approaches to Children's Literature,
https://doi.org/10.1007/978-3-030-52527-9_16

and America, with the hope that this may encourage further research on Latin translation and its relationship to children's literature and education in a global sense. In addition, I will consider how trends in Latin education in the English-speaking world have influenced the production of Latin translation in English children's literature. Finally, this analysis will consider the target readership and translational intent of these texts, and speculate on the future of Latin children's literature commercially, culturally, and educationally.

From its time as the *lingua franca* of the Roman Empire, Latin has retained a privileged status in Western scholarship and culture. While it would decline as a spoken/vernacular language leading up to the fall of Rome, it would long stand as the written language of educational, ecclesiastical, and political matters throughout the Middle Ages and into the Renaissance. With the establishment of the earliest universities in the eleventh century, Latin grammar became the most foundational subject of the trivium (which also included logic and rhetoric), with Latin or "grammar" schools designated to instruct young children in the basics of the language. Czech philosopher and theologian John Amos Comenius, considered by many the father of modern education, stressed a balanced and sensory approach to children's learning of Latin that combined traditional written grammar studies with oral vernacular and pictorial examples (Thut 233). His *Orbis Sensualium Pictus*—generally regarded as the first modern picture book—was published in 1658 with the Latin and German texts alongside each other. (It was translated into a Latin-and-English text the following year by Charles Hoole.)

By the mid-nineteenth century, university instruction was gradually shifting from Latin to the vernacular in several Western nations, but familiarity with and childhood instruction in Latin was still near-universal among the educated. Charles Lutwidge Dodgson, who would write *Alice's Adventures in Wonderland* (1865) and *Through the Looking-Glass* (1871), enjoyed an ongoing fascination with Latin, including his well-known 1888 poem, "A Lesson in Latin," which is equal parts studious and whimsical ("Our Latin lesson is complete: / We've learned that Love is Bitter-Sweet!"). Dodgson, a long-time lecturer at Christ Church College at Oxford, in fact owes his literary pseudonym to his proclivity for the language of scholarship. Dodgson translated his middle and first names into Latin (Ludovicus and Carolus, respectively) and then anglicized them to create Lewis Carroll, perhaps the most internationally famous name in children's literature over the past century-and-a-half.

Within weeks of the original publication of *Through the Looking-Glass*, there were several Latin translations made of Carroll's "Jabberwocky" poem, with numerous other translations to follow in the decades to come. The two most famous translations would ultimately be those of Carroll's paternal uncle, Hassard Dodgson ("Gaberbocchus" in 1872), and Augustus Vansittart, a noted Biblical scholar at Trinity College, Cambridge ("Mors Iabrochii" in 1881). Hassard Dodgson stands as a rarity in the Latin translation of children's literature—an individual with a primary interest in the latter subject instead of the former—while Vansittart's classical academic background would become the model for the field. Rather than liberalizing Latin translation, children's literature has generally demanded an even higher set of qualifications for such translation than many other forms of literature; consequently, in contrast to Lawrence Venuti's suggestion of the invisibility of the translator, the translators of Latin children's literature have tended to be highly prominent, and critical conversation has often focused as much on the individual translators as the translations themselves.

While Carroll's verse proved a consistent object of intrigue for classical scholars and translators, there were no full-length Latin translations produced of either *Alice* book in the nineteenth century, and only a narrow selection of adventure novels would be translated over the next fifty years in the United Kingdom and America. Daniel Defoe's *Robinson Crusoe* was translated into Latin by both Francis William Newman (*Rebilius Cruso*) in 1884 and G. F. Goffeaux (*The Story of Robinson Crusoe in Latin*) in 1907. Meanwhile, Arcadius Avellanus translated Robert Louis Stevenson's *Treasure Island* (*Insula Thesauraria*) in 1922 and produced yet another translation of *Robinson Crusoe* (*Vita Discriminaque Robinsonis Crusoei*) in 1928.

By the end of the 1920s, Latin enrollment exceeded all other foreign language enrollments combined in US secondary schools, and the frequent translations of a popular schoolboy story like Defoe's novel are indicative of the pedagogical intent of these texts. In the preface to his translation, Newman (the brother of Cardinal John Henry Newman) stresses, "No accuracy of reading small portions of Latin will ever be so effective as extensive reading; and to make extensive reading possible to the many, the style ought to be very easy and the matter attractive" (v).[1] The following three decades would be highlighted by a more international trend in Latin translation for children. Wilhelm Busch's *Max und Moritz* was translated into Latin by Erwin Steindl in 1925 and again by

Gotthold Merten in 1932, and Henry Maffacini's 1950 translation of the Italian *Le avventure di Pinocchio* (*Pinoculus*) was joined by no fewer than five separate translations of the German *Der Struwwelpeter* from 1934 to 1960. Meanwhile, the first International Conference for Living Latin (officially titled *Congrès international pour le Latin vivant*) took place in Avignon in 1956, stressing the practical usage of the language and its relevance to contemporary texts.

In the midst of the international Living Latin movement, Alexander Lenard's 1960 translation of A. A. Milne's *Winnie-the-Pooh*—*Winnie Ille Pu*—would prove to be a paradigmatic event in the history of Latin translation of children's literature. Lenard was a Hungarian-born refugee physician living in Brazil who spoke twelve languages and had done work in the Vatican library during World War II, and his translation would become an unexpected cultural and commercial phenomenon. *Winnie Ille Pu* would spend twenty weeks on the *New York Times* best-seller list and to this day remains the only Latin text ever to appear on that list. Its success was heralded by a flurry of high-profile critical reviews: *The Christian Science Monitor* declared, "Even Caesar never took a country as large as America in two months' time"; the *New York Times* called it "the greatest book a dead language has ever known"; and *TIME Magazine* wryly termed it "a Latinist's delight, the very book that dozens of Americans, possibly even 50, have been waiting for" (McDowell).

While the image on the book's front cover of Pooh as a Roman Centurion implies that this translation is in the classical Latin of the Roman Empire, Lenard is in fact culling from five centuries of Latin verse from the Middle Ages and the Renaissance. This is often the case for Latin translators of children's literature, as classical Latin verse generally does not rhyme and utilizes a quantity-based (rather than stress-based) prosody. With the gradual transition to medieval, ecclesiastical, and Renaissance Latin, verse in Latin would more closely reflect contemporary English poetic conventions, and the predominant influence of these later Latin periods is evident in many of the texts under analysis here. As a case in point, in translating Milne's nonsense poem, "Lines Written by a Bear of Very Little Brain," Lenard opens with the line "Dies ille, dies Lunae" (71)—an obvious allusion to the well-known thirteenth-century Latin hymn "Dies irae," with the opening line "Dies irae, dies illa" (There is here also the ironic gesture of aligning Milne's light-hearted verse with a medieval hymn whose title translates as "Day of Wrath").

The critical and commercial success of *Winnie Ille Pu* helped spur further production in the field, and it was followed in the next five years by notable Latin translations of Munro Leaf's *Ferdinand the Bull* (*Ferdinandus Taurus*), Beatrix Potter's *The Tale of Peter Rabbit* (*Fabula de Petro Cuniculo*), and Antoine de Saint-Exupéry's *Le Petit Prince* (*Regulus*). However, it would be Clive Carruthers' Latin translations of both *Alice's Adventures in Wonderland* (*Alicia in Terra Mirabili* [1964]) and *Through the Looking-Glass* (*Aliciae Per Speculum Transitus* [1966]) that would become the second truly notable achievement in the field that decade. In producing the first full-length Latin translations of Carroll's seminal texts, Carruthers would boldly confront concepts of linguistic equivalency, using his previous work in translating the Old Latin comedies of Plautus as a model for recreating Carroll's nonsense stories and verse in Neo-Latin. The acrostic poem on *Alicia in Terra Mirabili's* back cover provides an instructive example of this, emulating Carroll's famous acrostic poem, which concludes the *Alice* books by invoking her name (the accordant English translation is presented on the right):

Animum adverte huic fabellae;	Pay heed to this little tale;
Licet scire mores hominum,	you may learn of the characters of human beings,
Introspectos oculis puellae	as viewed through the eyes of a little girl
Cui evenit mirum somnium,	who had a marvellous dream,
Indoles ineptas, ioca, gerras: —	their silly dispositions, jokes and nonsense: —
Aptum vitae hic compendium.	Here you have a fitting summary of life.
IN TERRA MIRABILI	**IN WONDERLAND**

Despite the success of a number of Latin children's texts in the 1960s—and despite the initial momentum of the Living Latin Movement—Carruthers' books represent a veritable last stand for this brief golden age of Latin translation in children's literature. The decline of Latin studies and usage in the English-speaking world would be both sudden and significant. The Second Vatican Council, which concluded in December 1965, formally endorsed the celebration of Catholic Mass in local vernaculars instead of Latin, thereby depriving the language of its most reliable and expansive forum. The University of Oxford and the University of Cambridge (institutions where instruction was once performed exclusively in Latin) had already both dropped their O-level Latin requirements for undergraduate admission in 1960 (Forrest 42), while Latin enrollment in American secondary schools—which had peaked in 1962—dropped seventy-nine percent in just over a decade (Kitchell).

Not surprisingly, Latin translation in children's literature also experienced a precipitous decline over the next few decades. Brian Staples' translation of Milne's *The House at Pooh Corner* (*Domus Anguli Puensis* [1980]), Bernard Fox's translation of E. B. White's *Charlotte's Web* (*Tela Charlottae* [1985]), and C. J. Hinke and George Van Buren's translation of L. Frank Baum's *The Wonderful Wizard of Oz* (*The Classical Wizard: Magus Mirabilis in Oz* [1987]) represent the most prominent translations of English titles to emerge in the next thirty years, although none of these texts would generate significant publicity or sales. Instead, the majority of notable Latin translations of children's texts in this time period would be Franco-Belgian comics, including Hergé's *Tintin*, Jacques Martin's *Alix*, and René Goscinny's *Asterix*—the latter of which had twenty-five volumes translated by Carolus Rubricastellanus, popular texts that also earned praise for their contributions to language learning.

After over three decades of relative public indifference, the 1998 initiation of the first large-scale publishing series of Latin children's translations would prove to be a paradigm-shifting event. Founded in 1979 by Ladislaus and Marie Bolchazy, Bolchazy-Carducci Publishers focused from its inception on Latin and Greek publications and instructional materials, but by the late 1990s its founders became interested in pursuing Latin translations of the verse of Dr. Seuss. Admittedly, the first published Latin version of Seuss (the highest-selling American children's author of the twentieth century) was the 1994 translation of *Oh, the Places You'll Go!*—*O, Loca Tu Ibis!*—by Leone Roselle for the J. Weston Walch educational publishing house. However, *O, Loca Tu Ibis!* is much more geared toward instruction than entertainment, as evident in its subtitle: *A Beginning Latin Reader and Activity Text*. It consequently bears little resemblance to Seuss's original book (the last of his storied career) beyond the theme and the translated segments of verse. The Latin version does not utilize any of Seuss's original illustrations and features student exercises at work's end accompanied by a separate teacher's guide. The vocabulary and grammar, while meant for Latin beginners, are intended to be learned in a classroom setting with teacher supervision, running somewhat contrary to Seuss's longstanding desire to make language learning a freeform and independent act.

Despite also being an educational publishing house by nature, Bolchazy-Carducci would resolve to have their translations better replicate the appearance and feel of the original Seuss books while also—in spite of the playful reputation of Seuss's illustration and verse—promoting Latin with technical virtuosity. The publishers enlisted as translators Terence and Jennifer Tunberg, an acclaimed pair of classics professors who would go on to found the Institute for Latin Studies at the University of Kentucky in 2001, currently the foremost spoken Latin program in American higher education. Their translation of *How the Grinch Stole Christmas!* (*Quomodo Invidiosulus Nomine Grinchus Christi Natalem Abrogaverit*) was published in 1998, and was followed by Latin translations of *The Cat in the Hat* (*Cattus Petasatus*) in 2000 and *Green Eggs and Ham* (*Virent Ova! Viret Perna!!*) in 2003. Save for the Latin wording and the Latin–English glossary at the end, these books appear virtually identical to Seuss's original texts—including all of the author's original artwork—with high-quality production values. In contrast to the lukewarm reception of most Latin children's literature since the mid-1960s, each of Bolchazy-Carducci's Seuss translations garnered publicity in high-profile publications: *Grinchus* appeared on the front cover of the *Wall Street Journal*, *Cattus Petasatus* earned a half-page editorial in the *Chronicle of Higher Education*, and *Virent Ova! Viret Perna!!* was prominently reviewed in the *Washington Post*.

With the initial success of the Seuss translations, Shel Silverstein's *The Giving Tree* was also translated by the Tunbergs as *Arbor Alma* in 2002, while Richard LaFleur—the contemporary author of the well-known *Wheelock's Latin* textbook—was selected to translate Maurice Sendak's *Where the Wild Things Are* as *Ubi Fera Sunt* in 2015.[2] While Bolchazy-Carducci's focus is on classical readers and instructional materials, their intended market for these children's books has been not only classicists but also consumers who normally would not buy Latin books—including those who want interesting coffee table books or avid Dr. Seuss collectors ("I suspect that a lot of them never read the book or even tried to," admits Marie Bolchazy). The popularity of the source texts was a huge factor in their selection, with the aim of these translations being—in Bolchazy's words—"the responsible popularization of Latin." This offers an unusual application of Venuti's insistence that "the aim of translation is to bring back a cultural other as the same, the recognizable, even the familiar" (18). Unlike in many works of translation, where an unfamiliar foreign text is familiarized by a vernacular language, in the case of Latin translation it is often the language itself that represents the cultural other for the target reader.

The efforts of Bolchazy-Carducci to popularize Latin through children's books have not been without complication. For example, when the Tunbergs decided to use "Invidiosulus" ("envious little wretch") as their translation of the word "Grinch," Dr. Seuss Enterprises would not grant permission unless the main character's original name was kept in the title, and the ensuing dispute held up production for six months.[3] In the end, the Tunbergs compromised with "Grinchus" while also keeping "Invidiosulus" on the front cover, and the resulting title translates somewhat bulkily as "How an envious little wretch, Grinch by name, stole the birthday of Christ" (Reardon 67).[4] Such negotiation applies usefully to Riitta Oittinen's emphasis that fidelity to the reader of the target text should supersede fidelity to the source text, but also to her observation that "the audiences of children's books may change in translation" (36). By the admission of both translator and publisher, *Grinchus* is a book that represents a highly challenging (if near impossible) read for small children. "We tried to create a fun Latin text," Terence Tunberg explains, but "the Latin is not baby Latin. The Latin is full-fledged Latin. It's not designed for beginning students" (Reardon 71). Meanwhile, Bolchazy stresses that their linguistic commitment is to the classicists, and if given the choice they would not have these texts stocked in children's book sections. As such, the question emerges whether these books offer a dual address to adults and children, or rather a single address to adults that is disguised as children's literature. To put it more plainly, if such texts are not being translated for children, do they cease to be children's literature?

Despite their prominent results translating picture books to Latin, Bolchazy-Carducci ultimately turned down the opportunity to produce full-length translations of J. K. Rowling's Harry Potter series (Bolchazy). Instead, Bloomsbury would commission Peter Needham—a Latin and Greek instructor at Eton for over thirty years—to translate the first two texts in the series: *Harrius Potter et Philosophi Lapis* (2003) and *Harrius Potter et Camera Secretorum* (2006). (Needham also translated Michael Bond's *A Bear Called Paddington* [*Ursus Nomine Paddington*] into Latin in 1999.) These translations have enjoyed the enthusiastic support of Rowling herself, an initial classics major at Exeter who liberally utilizes her earlier study in the original Harry Potter books—from an extended series of Latin spells to the Latin motto of Hogwarts School of Witchcraft and Wizardry: "Draco Dormiens Nunquam Titillandus" ("Never Tickle a Sleeping Dragon"). As a result, despite its comparative bulk the Harry Potter series often offers a clearer application of classical Latin linguistics than many of the previously mentioned translations. Needham's choice of "Harrius" as the translation of "Harry," for example, stems from the "Arrius" name used by the Roman poet Catullus in a first-century-B.C. poem—a humorous elegiac couplet in which the subject (Arrius) insists on putting an "h" sound in front of words, contrary to classical Latin pronunciation.

While most of the aforementioned titles translated into Latin form a veritable high canon of children's literature, there are seeming exceptions to this trend, as is the case with Robert Dobbin's 2004 Latin translation of William Kotzwinkle and Glenn Murray's *Walter the Farting Dog*. It is easy to suggest that Dobbin's *Walter Canis Inflatus* was produced purely for cultural shock value—a lowbrow story heavily reliant on scatology for both humor and narrative translated into the historic language of intellect. However, to do so ignores the similarities of *Walter Canis Inflatus* with the majority of previously mentioned titles: It stems from a proven commercial commodity (the English original reached the top of the *New York Times* children's best-seller list) and its translator is a distinguished classicist (Dobbin would later translate the works of Epictetus for Oxford University Press). It is also one of the very few Latin children's translations that offer the original English text on the same page as the Latin translation, and online reviews of the book indicate that a number of children are reading the text with their parents. As such, Dobbin's *Walter Canis Inflatus* is likely (and ironically) the modern children's book in closest alignment with Comenius's *Orbis Sensualium Pictus* and its original employment of Latin alongside vernacular and pictures.

After the flurry of Latin translation of children's literature in the late 1990s and early 2000s, production has been steadily incremental over the past decade-plus. New Latin translations have been produced nearly every year, highlighted by Amy High's translation of Ian Falconer's *Olivia* (*Olivia: The Essential Latin Edition*) in 2007, Mark Walker's translation of J. R. R. Tolkien's *The Hobbit* (*Hobbitus Ille*) in 2012, and Monsignor Daniel Gallagher's translation of Jeff Kinney's *Diary of a Wimpy Kid* (*Commentarii de Inepto Puero*) in 2015. As is the case with Lenard's *Winnie Ille Pu* and several other Latin titles, both High's and Gallagher's translations feature their main characters in Roman garb on the front cover. Gallagher's title also provides a clear allusion to Julius Caesar's first-century-BC narrative of the Gallic Wars, *Commentarii de Bello Gallico*, which is also one of the most commonly utilized primary texts in contemporary Latin education. Hence, whereas Göte Klingberg touches on the challenges of the modernization of the classics (56–57), contemporary translation into Latin is in effect an effort to classicize the modern. Like the Latin translations before them, these works feature eminent and high-profile translators: *Commentarii de Inepto Puero's* translator, Monsignor Gallagher, notably served as the Papal Latin Secretary for the Vatican for eight years under both Pope Benedict XVI and Pope Francis.[5] As with many of the Latin translations of children's literature that have preceded them, the general public's engagement is with a familiar text in an unfamiliar language. In an opening note of appreciation to Gallagher, Kinney writes that he hopes this translation will "bring *Diary of a Wimpy Kid* to life in a way that will help people all over the world gain a deeper appreciation of this wonderful, and vital, language."

In the wake of this general resurgence in Latin translation of children's literature, a pair of situational perspectives provide useful commentary on the future of this trend. The first comes from Ben Harris, a former classics editor for Cambridge University Press who translated Julia Donaldson's *The Gruffalo* into Latin in 2012. Contrary to being recruited by a publisher for the task, Harris undertook the translation at the suggestion of his brother out of admiration for the work of Donaldson (a personal friend) and a love of Latin. While Donaldson was highly supportive from the outset, securing the rights from the publisher was an arduous task given the almost negligible commercial value for the company. With very few exceptions, Latin children's literature is a largely pro bono exercise for both publisher and translator—more akin to the ethical dynamic of esoteric academic publishing—with the motivation consistently being the promotion and/or canonization of the author/work. Simply put, says Harris, "I figured that if *Winnie-the-Pooh* was worthy of translation into Latin, then so was *The Gruffalo*."

Despite translating a book originally intended for young children, Harris mirrors the Tunbergs in not using children as the target readers for the Latin translation. While his recognition of the joys of children's literature is obvious, Harris pragmatically admits that the ideal learning environment for Latin should involve a qualified language teacher and instructional materials that emphasize authentic cultural context. The Latin edition of *The Gruffalo* consequently eschews a beginner mentality;

references to Ovid's *Metamorphoses* and Virgil's *Aeneid* are evident throughout the text, and Harris emphasizes that the book is "really a kind of clever conceit for knowing adults" more than a practical language primer for children. The element of nostalgia so present in many intergenerational picture book readings is in this case provided by Latin itself—for "the parents and grandparents," Harris says, who "would have remembered their Latin from school with fondness and would enjoy revisiting it through an iconic modern picture book."

The second perspective on the future of Latin translation in children's literature comes from Lucianne Junker, a protégé of the Tunbergs who has pushed for the accessibility of Latin for the very young. Despite the plethora of recent Latin children's books, Junker expresses frustration at the lack of Latin board books that children four years old and younger can engage with, stressing a positive preliterate introduction to the language that is specifically intended for a child target reader. With this in mind, she has independently translated a pair of celebrated board books into Latin—Bill Martin, Jr.'s *Brown Bear, Brown Bear, What Do You See?* [*Urse Fusce, Tueris Quem?*] and Eric Carle's *The Very Hungry Caterpillar* [*Eruca Vorax*]—literally gluing the translations over the original words of the cardboard books to read with her own children. Despite the virtuosic quality of her translations of two of the best-selling board books in American history aimed at a largely untapped market, she has initially encountered tepid interest from publishers, perhaps underscoring the general understanding that Latin translation of children's books must target adult readers instead of children.

This is unfortunate, as even (indeed, especially) something as simple as a board book intended for preschoolers stresses the highly intricate technical competency and literary comprehensiveness required for effective contemporary Latin translation. In *Urse Fusce, Tueris Quem?*, colors without a classical equivalent offer a persistent challenge, as does the fact that Latin words for animals and colors are generally longer (metrically) than their English equivalents. In the case of the color purple, Junker avoided the common contemporary translation of the word ("purpureus," which actually means something more like maroon) in favor of "ostrina" for its syllabic concision, its accurate description, and its historic usage by the Augustan elegiac poet Propertius. Meanwhile, in *Eruca Vorax*, multiple nods to Horace, Erasmus, and other luminaries of the Latin canon are utilized, a classical acumen that extends even to the title. While her original translated title of *The Very Hungry Caterpillar* was the syllabic-equivalent *Eruca Valde Esuriens* [The Caterpillar Hungering Very Much], Junker eventually shifted it to the more concise *Eruca Vorax* based on Cicero's memorable usage of the word "vorax" to describe the all-consuming mythical whirlpool Charybdis. In doing so, she emphasizes the great care and depth accorded to all great works of children's literature, elevating the popular form with an intellectual language intended for introduction more than instruction. At a time when the market for Latin translation of children's literature has somewhat plateaued, Junker has highlighted a practical lack that may portend the future direction of production in the field.

In his second-century-A.D. *Satires*, the Roman poet Juvenal writes that "maxima debetur puero reverentia" ["the greatest reverence is due the young"] (XIV:47), a statement which proves just as complex today when reflecting on children's literature in Latin. Despite the global aspirations evident in Kinney's aforementioned opening note to *Commentarii de Inepto Puero*, there remain a number of questions and challenges to consider for gauging the long-term prospects for the Latin translation of children's literature. According to Emer O'Sullivan, "Comparative children's literature questions the system of children's literature, its structure of communication and the economic, social, and cultural conditions that allow it to develop" (190). With this in mind, one can reasonably ask whether bringing children into contact with Latin is equivalent to contact with other cultures, or whether it simply provides a glorified linguistic exercise. Slightly more cynical is the potential suggestion that Latin translation of children's literature is an unnecessary pedantic exercise that is heavy on critical intrigue and low on practicality for children—in other words, that it is novelty literature instead of comparative literature. Or worse, that it is simply a joke to begin with, and that Lenard's *Winnie Ille Pu* is more akin to Frederick Crews's 2001 *Postmodern Pooh*—a humorous parody of academia and criticism—than to Milne's original *Winnie-the-Pooh*. Most extreme is the question of whether the current Latin translation of children's literature is actually an *anti*-comparative gesture, given that its source texts are generally well-established titles in English that have already proven commercially profitable. Rather than diversifying the children's literature canon, does Latin translation simply reaffirm its existent borders?

In spite of this, there are a number of equally compelling reasons to seriously contemplate and encourage the confluence of Latin and children's literature. To begin with, it offers a complex negotiation of the high/low and elite/popular dynamics that have historically separated Latin and children's literature. Perhaps the humor that many find in such juxtaposition is simply indicative of the continuing intellectual and academic marginalization of children's literature. However, much as Latin holds the potential to augment the scholarly legitimacy of children's literature, the use of Latin in children's literature offers the language a popular relevancy it often struggles for in the contemporary world. In addition, it is crucial to address the ways that Latin translation usefully complicates existing translation methods and objectives for children's literature, as Bolchazy's stated goal of the "responsible popularization" of language rather than text makes evident. Furthermore, even if these texts are geared toward adults, they may still be responsible for stoking an interest in children and for encouraging communal language learning within the family. If so, it is worth considering the cultural and intellectual mediation that Latin offers between adult and child readers, making the elite accessible and the mundane enjoyable for both potentially uninitiated age groups. Viewed in this manner, the contemporary rise of Latin translation in children's literature may herald a comparative rebirth of classical education and influence for all ages in the English-speaking world, suggesting the possibility of more widespread distribution of these texts as educational material. At its most dynamic, such mediation affirms the contemporary relevance of both Latin and children's literature—a language ostensibly situated in the past and a literature presumably oriented toward the future—with a translational overlap that is as potentially significant and productive as it is surprising.

Acknowledgements My sincere thanks to Micah Phipps for his vital role in the origination and development of this project. This was first presented at the Translation Studies and Children's Literature symposium in Brussels and Antwerp in October 2017, and my gratitude to Vanessa Joosen, Clémentine Beauvais, and Emer O'Sullivan for their feedback and suggestions for finalizing the article appearing in *Bookbird*. It was presented in expanded form at the 2018 Neo-Latin Symposium in Lexington, Kentucky, and my appreciation to the collective group of symposium participants for their enthusiastic response and invaluable insight in bringing this expanded chapter to fruition.

Notes

1. This is also made evident in Newman's subtitle for *Rebilius Cruso*: *A Book to Lighten Tedium to a Learner*.
2. Bolchazy-Carducci's children's book offerings also include a series of original animal stories in Latin by Rose R. Williams and a set of original readers by Marie Bolchazy (and translated by Mardah Weinfield) focusing on linguistic comprehension of numbers, food, and colors in Latin. Meanwhile, Terence Tunberg has elsewhere collaborated with colleague Milena Minkova on a Latin translation of Mother Goose verse: *Mater Anserina: Poems in Latin for Children*, Focus Publishing, 2007.
3. Of course, such complication and controversy regarding names is in no way limited to Latin translation. For a more detailed analysis of the politics of this process, see Jan Van Coillie's "Character Names in Translation: A Functional Approach," in *Children's Literature in Translation: Challenges and Strategies*, St. Jerome Publishing, 2006, pp. 123–139 and Yvonne Bertills' *Beyond Identification: Proper Names in Children's Literature*, Åbo Akademi University Press, 2003.
4. This title represents a minor aberration in the collective quality and cohesion of the Bolchazy-Carducci Seuss books, as the Tunbergs' translations are to be roundly commended for their precision, creativity, and (from a Latinist's perspective) entertainment value.
5. In 2017, Gallagher left the Vatican for Cornell University, where he is currently the Ralph and Jeanne Kanders Associate Professor of the Practice in Latin.

Works Cited

Children's Books (in Alphabetical Order by Translator)

Carruthers, Clive Harcourt, translator. *Alicia in Terra Mirabili*. By Lewis Carroll, St. Martin's, 1964.

Dobbin, Robert, translator. *Walter Canis Inflatus*. By William Kotzwinkle and Glenn Murray, Frog, Ltd., 2004.

Gallagher, Daniel B., translator. *Commentarii de Inepto Puero*. By Jeff Kinney, Amulet Books, 2015.

Lenard, Alexander, translator. *Winnie Ille Pu*. By A. A. Milne, E. P. Dutton, 1960.

Newman, Francis William, translator. *Rebilius Cruso*. By Daniel Defoe, Trübner & Co., 1884.

Tunberg, Jennifer, and Terence Tunberg, translators. *Quomodo Invidiosulus Nomine Grinchus Christi Natalem Abrogaverit*. By Dr. Seuss, Bolchazy-Carducci, 1998.

Secondary Sources

Bolchazy, Marie. Personal Interview. 13 October 2017.

Forrest, Martin. "The Abolition of Compulsory Latin and Its Consequences." *Greece & Rome*, vol. 50 (The Classical Association: The First Century 1903–2003), 2003, pp. 42–66.

Harris, Ben. Personal Interview. 30 June 2019.

Junker, Lucianne. Personal Interview. 23 November 2018.

Kitchell, Kenneth. "Teaching of Latin in Schools: Enrollments, Teaching Methods and Textbooks, Issues Trends and Controversies." *Stateuniversity.com*, education.stateuniversity.com/pages/2160/Latin-in-Schools-Teaching.html.

Klingberg, Göte. *Children's Fiction in the Hands of the Translators*. CWK Gleerup, 1986.

McDowell, Edwin. "'Winnie Ille Pu' Nearly XXV Years Later." *The New York Times*, 18 November 1984, nytimes.com/1984/11/18/books/winnie-ille-pu-nearly-xxv-years-later.html.

Mead, Rebecca. "Eat, Pray, Latin." *The New Yorker*, 20 July 2016, newyorker.com/culture/cultural-comment/eat-pray-latin.

Oittinen, Riitta. "No Innocent Act: On the Ethics of Translating for Children." *Children's Literature in Translation: Challenges and Strategies*, edited by Jan Van Coillie and Walter P. Verschueren. St. Jerome Publishing, 2006, pp. 35–45.

O'Sullivan, Emer. "Comparative Children's Literature." *PMLA*, vol. 126, no. 1, 2011, pp. 189–196.

Reardon, Patrick T. "How the Grinch Went Latin: Or, How 'Invidiosulus Nomine Grinchus' Got Translated into an Ancient Language to the Great Merriment of All." *Chicago Tribune*, 15 December 1998, pp. 67+.

Thut, Isaak Noah. *The Story of Education: Philosophical and Historical Foundations*. McGraw-Hill, 1957.

Venuti, Lawrence. *The Translator's Invisibility: A History of Translation*. Routledge, 1995.

Newtonian and Quantum Physics for Babies: A Quirky Gimmick for Adults or Pre-science for Toddlers?

Casey D. Gailey

Wandering the bookstore one weekend, I stopped in my tracks upon spotting a board book by the children's section—not because the cover was particularly alluring or exciting but because the title, *Quantum Physics for Babies*, stunned me. As a biologist, I recall working through a college physics class, struggling through the details of quantization and excitation of electrons. Why on earth was there such a book in the children's literature section? This astonishment and curiosity propelled me into buying the book and a few others—such as *General Relativity for Babies*, *Quantum Entanglement for Babies*, *Baby Loves Thermodynamics!*, and *Baby Loves Aerospace Engineering!*. Were these quirky books actually for toddlers or for the parodic amusement of science-minded adults?

Board books are familiar toys for babies and toddlers. The hardy, non-toxic book is safe in the hands of babies and the illustration of basic ideas is a great learning tool. These board books help reinforce object identification of toys and animals, as well as concepts like bedtime and playing. Some board books can transcend utilitarianism and become pre-literature rather than merely learning toys. According to Kent State University

C. D. Gailey (✉)
Vanderbilt University in Nashville, Nashville, Tennessee, USA

A. Kérchy and B. Sundmark (eds.), *Translating and Transmediating Children's Literature*, Critical Approaches to Children's Literature,
https://doi.org/10.1007/978-3-030-52527-9_17

Literature scholar Marilyn Apseloff, in "Books for Babies: Learning Toys or Pre-literature?," some of these baby books can be called pre-literature because they "not only introduce a child to the format of a book but also to the world of the imagination, to narrative through verbalization, and to art through a variety of illustrations and styles" (63). This allows for a framework from which to study these Science for Babies books. Just as some board books can be considered pre-literature, working as a precursor to more traditional children's books, perhaps these could provide pre-science to the youngest audience.

Science is present in children's literature, of course, but this trend for science board books is a recent movement. This is likely targeting the recent generations of adults who were progressively encouraged toward science, technology, medicine, and other areas of study due to the focus on STEM in the last decades. However, science is not just for adults. As a science writer and an editor Patricia Lauber pointed out in her article "What Makes an Appealing and Readable Science Book?," children are "born curious, wanting and needing to understand the world around them, wanting to know why, how, and what: the very questions that scientists ask" (5). Science has a natural appeal to children, to their powers of reasoning and questioning, their intuition and imagination. In this way, children's science books can serve a pedagogical purpose, whether they are informational or narrative books, by targeting that innate curiosity and attempting to cultivate it further.

Moreover, such science books can serve another pedagogical purpose in their ability to express scientific ideas in a way to accentuate scientific education. In "Children's Literature and the Science Classroom," Sandra K. Abell, former elementary teacher and Professor and Director of the University of Missouri Science Education Center, explains that children's literature in both "narrative (story based) and expository (informational) forms" can "address many scientific topics" because they can "take children to places that they could not go on their own and allow them to explore natural phenomena that might be too small or take too long to observe directly in the classroom" (54). Although the target audience Abell's article focuses on is that of elementary-school-aged children, the same concept can apply to these Science for Babies board books. For instance, Ferrie's *Quantum Physics for Babies* book describes atoms and their particles, something which is too miniscule to be observed, but books can transcend scale and zoom into the atomic level in order to explain the concept of atoms. However, this raises the issue of the balance between informational accuracy for the sake of pre-science and accounting for the translation and condensation of concepts when transmediating scientific principles into children's literature. This enables an alternative way to analyze children's science books: Instead of analyzing such books merely from the context of education or marketing, it is possible to consider such books as adaptations.

Adaptation theory is well known in the realm of young adult and children's literature. The current frame of mind regarding adaptions is that, while they are inherently "haunted at all times by their adapted texts," which enables a palimpsestic pleasure, adaptions must also have individual worth, which differs from but is no less valuable than its originary source (Hutcheon 6). Regardless, the transtextuality between the urtext and adaptations is a vital area of study. In Robert Stam's article "Beyond Fidelity: The Dialogics of Adaptation," he discusses Gérard Genette's ideas of transtextuality. One form of intertextuality is that of "hypertextuality," in which one text "transforms, modifies, elaborates, or extends" an earlier text, the "hypotext" (65). In this way, one can analyze the intertextual relationships of literature to other books, movies, paraphernalia, TV, fanfiction, and more. However, books such as the Baby Loves Science! and Science for Babies series stray from this traditional situation. These board books work in a similar way to other adaptations, in that they translate and remold the originary "text" into a new format, with a new target audience; however, these adaptations are not necessarily derived from a particular hypotext, not from literature or popular media, but rather from collective scientific theory. In this way, the science board books are adaptations, but they are not as restricted into form, plot, or style as traditional adaptations are in their nature of reflecting a more distinctive hypotext.

When attempting to critique the nature and adaptive value of the Baby Loves Science! and Science for Babies series, one could lean toward three areas of thought: (1) that the texts are intended and presented as parodic and are more a novelty for adults who understand the originary scientific theories presented; (2) that the texts are expository and seek to express the complex scientific topics to young children; or (3) that the texts do not assume to truly relate the theories, but rather to inspire the target audience to seek out science as they age.

There is certainly an essence of parody in these baby science books. The juxtaposing of complex scientific theory such as quantum physics, which many adults struggle to grasp, with the simplicity and puerility of a baby's board book is humorous in its absurdity. However, as Lauber argues, children's books are not generally chosen by the children, especially baby books, but rather are selected and bought by the adults (5). Thus, this veneer of parody is useful as a marketing ploy for appealing to the adults in order to reach children. This sense of humor enables the books more success whereas as an ostensibly didactic book would be repellent to the adult and child.

Despite the somewhat parodic nature of these books, the goal of this paper is to consider three board books as pre-science: *Newtonian Physics for Babies* and *Quantum Physics for Babies* by Chris Ferrie and *Baby Loves Quantum Physics!* by Ruth Spiro. I argue that based on considerations of the cognitive requirements of science, use of picturebooks in acquisition of literacy, analysis of visual and linguistic design elements, accuracy of information, and the pre-science and future-looking potential of these books the ostensible purpose lies in encouraging young readers toward science as they mature.

Cognitive Requirements of Science

In her article "Is There Any Such Thing as Children's Science Fiction?: A Position Piece," Farah Mendlesohn argues that while younger children may be interested in science fiction, "the more we considered the ideological and cognitive demands of science fiction, the more we became uncertain as to whether this mode of writing can be adapted wholesale for a younger audience" (285). As curious as children may be about science and science fiction, there are cognitive requirements of science fiction that create cognitive dissonance (285). Namely, Mendlesohn's article dissected out the struggle that science fiction is speculative, but in order to create "what ifs," the readers need the cognitive groundwork in order to build off of. Mendlesohn says that science fiction is more "a way of thinking about the world which requires authors to offer a type of challenge" but that this is contrary to the "assumption that literature for children should reinforce what they already recognize" (285), which arouses an "insistence on didacticism" because of "the expectations and understandings of children's cognitive and emotional development" (286). This struggle for writing science fiction is reflected in science books generally because authors must balance expressing scientific concepts while working on the premise that children might not have the cognition required to understand and that, as in any conventional Typical Case Prototype children's literature (as George Lakoff would say), children emotionally need reassurance that the universe is safe and stable.

But what is the assumed cognitive ability of toddlers? As an example, according to the Virginia Department of Education's Science Standards of Learning for kindergarten curriculum, kindergarten students are expected to learn the following introductory concepts: Matter and identifying physical properties of objects (colors, shapes, textures, and size), making observations, learning the five senses and describing sensation, learning about magnetism, identifying solid/liquid/gas, identifying living animals/plants vs. nonliving objects, learning basic life cycle and needs of plants/animals, understanding shadows, observing basic weather patterns, understanding how things/people can change over time, and learning about recycling (2–33). Based on this and the logical assumption that younger children, such as the toddler age group, would have less developed cognitive ability, it is unlikely that the targeted audience of the Science for Babies books would be fully intellectually capable of grasping such complex and abstract concepts as quantum physics. However, because kindergarten-aged children are expected to learn a slightly abstract concept like magnetism, one can argue that it is still possible that the target audience would be able to absorb at least some understanding of ideas like atoms making up all objects or the basic theory of gravity. Thus, this supports the potential of the books as a source of pre-science that instills basic groundwork and scientific thinking.

Use of Picturebooks in Literacy Acquisition

Apseloff provides five conditions for baby books to be classified as pre-literature: that the book (1) introduces children to the format of a book, (2) stimulates imagination, (3) provides "narrative through verbalization," (4) offers "art through a variety of illustrations," and (5) stimulates a relationship between the adult mediator and child (64).

Bettina Kümmerling-Meibauer and Jörg Meibauer, in their article "First Pictures, Early Concepts: Early Concept Books," suggest that toddlers do not understand books except perhaps as a teething aid and "need an adult or older child to learn the 'rules of book behaviour' ... [including] sitting still, turning the pages, looking and pointing at the picture" (340). They assert that in the process of learning book formatting, children must "learn that a book has a beginning and an end, often noted by the adult's interjection 'the end'" and learn the "sense of linear order" (340). Thus, in considering the first condition, one must assume the fifth condition is met—that an adult or older child is present to mediate the reading and that this adult uses the board book to introduce what a book is and how it is structured. The true requirement for this condition is that the board book has a linear progression with an identifiable start and end, which occurs with each page turned. This is opposed to some board books, which merely present random images like "apple," "dog," and "house" for simple object-word acquisition.

All three of these Science for Babies books meet this requirement. Both Ferrie's *Newtonian Physics for Babies* and *Quantum Physics for Babies* use a gradual process that starts with one idea and builds on to it with further detail. In *Newtonian Physics*, the book begins with the broad statement of "This is a ball" (1) developing into "The ball feels the force of gravity" (2) and properties like mass and acceleration until the books amount to "This is Sir Isaac Newton. He wrote three laws of motion" (18) and ending with "Now you know Newtonian Physics!" and a brief description of each law (22). Likewise, *Quantum Physics* follows this method again starting broad with "This is a ball" (1) to orient the audience before leading to "All balls are made of atoms" (5) and describing the parts of an atom, electrons, and what quantization is before finishing on "Now you are a QUANTUM PHYSICIST!" (22). Spiro's *Baby Loves Quantum Physics!* also has a gradual process and is more narrative than Ferrie's books, starting with introducing that "Baby loves Cat!" (1) with a picture of "Cat" (1) and "Baby" (2) before starting the story that "Sometimes Cat likes to hide" (3) and eventually reaching the cat hiding in a box and that "In quantum physics, until Baby looks in the box, Cat is both asleep *and* awake" (16).

For the second condition, all three books can arouse the imagination as they introduce new ideas and images to the audience who have presumably never been exposed to before. The third condition is also met as all three books have a narrative structure that uses text in addition to images to develop the story. The fourth condition is met in that the books all make use of a variety of images and artistic design elements to support the linguistic text.

Summarily, Spiro's book and both of Ferrie's books meet at least Apseloff's requirements to be classified as pre-literature.

Analysis of Visual Design Elements

In their article, Kummerling-Meibauer and Meibauer discuss the visual features of early concept books. They attempt to compile essential similarities of early concept books that explain the visual constraints of board books. According to their analysis, there are four "essential skills ... which must be learned like the rules of verbal language" (333). These are the ability to "distinguish a figure from the background" in order to "to realize that the depicted object is relevant" rather than the background; to realize that "that lines, points, and colors have meanings" and can do more than define objects from the background such as by suggesting textures or shade; to understand "that two-dimensional pictures stand for three-dimensional objects" that can be experienced; and to recognize similarities between "real and depicted objects" (332–333). In this way, Kummerling-Meibauer and Meibauer suggest that board books have essential visual design elements that assist in the acquisition of visual language in addition to verbal language. Moreover, that visual elements must be specifically chosen to enhance the clarity of objects presented in picturebooks and the audience's understanding and ability to recognize that the objects in the books represent real things.

Kummerling-Meibauer and Meibauer emphasized that early concept books must distinguish the object from the background in order to emphasize its relevance, as more important to focus on. All three of the Science for Babies books do this. In Spiro's book, the main points of focus for the narrative, Baby and Cat, are contrasted against a solid, light-tinted teal (1–2, 7, 17–18), light gray (3–6, 19–20), light green (8), light yellow (9–10), medium purple (11–12, 15), or dark orange (13–14, 16) background. Even on pages where Baby's nursery is depicted via a crib, nightstand, rug, and hanging picture, the three-dimensionality of the room is lost by the solid backdrop. Furthermore, as Ellen Lupton and Jennifer Cole Phillips explain in *Graphic Design The New Basics*, "Colors that sit near each other on the spectrum or close together on the color wheel are analogous. Using them together provides minimal colors contrast and an innate harmony" (82). From this, one could understand the furniture is colored in light hues of blue, green, white, gray, yellow, and pink, which have such minimal contrast due to their light hues, so that they provide a neutral backdrop; they assist a young audience in contextualizing the story because it provides a setting, but they direct the focus onto the images of Baby and the box Cat hides in, which both use the bolder orange and purple colors. Ferrie's books more strongly stress the focus on the objects instead of the background as, in both books, the background is always white, whereas the objects, often a ball being described by a characteristic such as feeling the force of gravity, are shown in bright colors like green, purple, and blue.

Lupton and Phillips also describe how images can be grouped based on "size, shape, color, proximity, and other factors" as well as how "As a process of separating, grouping serves to break down large, complex objects into smaller, simpler ones" (102). This provides an interesting basis to analyze the Spiro and Ferrie books. All three books intricately make use of color, shape, and size and a rhetorical approach to group concepts together and to break down large concepts. In Ferrie's *Quantum Physics* book, for example, in describing the parts of the atom, color is used to obviously link the vocabulary of "proton," "neutron," and "electron" to their correlation with the atom as a whole. This helps to piece apart the idea of an atom, which is presumably foreign to the audience, therefore simplifying it, while also grouping smaller ideas together in order to provide clarity and distinction. The word "neutrons" is written in the same color purple as the circles in the nucleus of the atom that illustrate neutrons, while the word "protons" is written in the same red that depicts the protons in the nucleus (6–7). The electrons are also shown this way, with the word written in the same green as the circles orbiting the atom (8). This is further maintained through the book so that, as new ideas are introduced, such as that electrons can exist in the electron rings but not elsewhere, the audience still has a frame of reference to recognize what the book means. Ferrie's *Newtonian Physics* also does this in the use of dark green downward arrows, which represent the downward force of gravity, and the repetition of coloring these arrows and the text "force of gravity" in the dark green throughout the book. This is contrasted with the upward arrows, which represent the equal and opposite force of the ground; both the arrows and text "force of the ground" are depicted in red. The choice to use the opposite and complementary red/green for the arrows of net force further helps link the relationship of the concepts. Spiro's Quantum Physics books also utilize color to group concepts. Namely, Spiro's book uses medium purple and dark orange, the only bold colors used in the book, to emphasize the dichotomy of Schrodinger's cat. On the box in which the cat hides, one face of the box is colored purple with a pink crescent moon and the other face shown is colored dark orange with a yellow sun. As the narrative reaches its climax, the same purple is used on the entire background accompanied by the text "Maybe Cat is asleep" and the orange is used as the background for "Maybe Cat is awake" to emphasize the paradoxical idea that the cat "is both asleep and awake" until the box is opened based on physics theory.

Based on this, the visual design elements are carefully selected and worked to emphasize the images and concepts expressed by the books—supporting the idea that the books do have some pedagogic implications since a merely parodic book would not focus on ways to actually elucidate the theories.

Analysis of Linguistic Design Elements

Spiro's book, *Baby Loves Quantum Physics!*, is narrative rather than expository. In order to present the idea of Schrödinger's cat, the book tells a simple story of Baby and Cat in which Cat is hiding and Baby searching for it.

Ferrie's books, *Quantum Physics for Babies* and *Newtonian Physics for Babies*, work on a different level than Spiro's. Ferrie's books are more expository in form, and they progress as a series of statements that build into larger concepts. For example, in introducing an atom, it starts with a ball, which is familiar, and moves into stating that "All balls are made of atoms" (5) before distinguishing neutrons, protons, and electrons (6–8).

The language, vocabulary, and syntax of the text used in these board books are dubious. For the majority of the books, the sentence structure is simplistic: short, direct statements directly associated with the images and using casual vocabulary. For example, in Spiro's *Baby Loves Quantum Physics!*, until the end of the board book, all of the sentences consist of five words or less, which are often depicted concretely by the illustrations. When the cat hides, the story says "Baby wants to find him!" and follows with "Is Cat here?" as Baby looks under the nightstand (7) and the "Or here?" as Baby looks in the crib (8). Even the decision to refer to the characters as "Baby" and "Cat" eliminates the need for a random name to associate with the characters and eliminates the definite article to make the sentences simpler to read. Thus, the books are on par with traditional baby books. However, as the text evolves into "In quantum physics, until Baby looks in the box, Cat is both asleep *and* awake" (16), the syntax becomes more complex and the familiarity is lost. This statement has now introduced "quantum physics," which is far beyond the knowledge of most children and articulating the concept that the cat can be awake and asleep at the same time. Additionally, the syntax has amounted to a sentence with three parts and includes a subordinate clause. This occurs with the next text "When Baby opens the box, she will find out if Cat is asleep or awake" (17), which is also a more complex structure than the rest of the book. While the target audience's presumed cognition may grasp the basic narrative, this ending suggests that, while the audience may understand some of the meaning, they are unlikely to fully grasp the concept expressed—supporting the hypothesis that this book functions more as pre-science than truly expository.

The Ferrie books are arguably more complex than the Spiro book. Although the syntax remains simple throughout *Quantum Physics* and only becomes slightly more complex at the end of *Newtonian Physics*, the vocabulary and concepts are more difficult. Spiro's book only really incorporates the scientific principle at the end, but Ferrie's books present new words throughout the board books. *Newtonian Physics* introduces the word "gravity" immediately and quickly follows with the concept of the force of gravity and its opposite, the force of the ground (2–7). In the following pages, the books use terms like "net force" (9), "accelerates" (11), and "mass" (12). *Quantum Physics* similarly uses terms like "energy" (2), "atoms" (5), "neutrons" (6), "protons" (7), "electrons" (8), and "quantized" (16). Considering that learning about mass was listed as a goal for kindergarteners in the Virginia Department of Education's curriculum, it is unlikely younger audiences could be familiar with any of these concepts before experiencing these series. Moreover, the books do not palpably explain some of those concepts, which suggests the audience should either know it or that it does not matter if the audience fully understands. For instance, although *Quantum Physics* provides "This is a ball" (1) and "All balls are made of atoms" (5), the text never pushes what an atom is. The book moves into naming the parts of an atom (neutrons, protons, and electrons), but skips saying what an atom itself is. Thus, the linguistic elements of these books even more than Spiro's are designed in a way that may impart some basic conceptualization of scientific principles to the audience, it is unlikely to be fully understood and so the books cannot be truly didactic or expository.

Accuracy of Scientific Information

Both series have authorial ethos, which contribute to the credibility and accuracy of the information in the books. Chris Ferrie's books are especially so since Ferrie, the author and illustrator, is a physicist, mathematician, and father. Neither Ruth Spiro, the author of *Baby Loves Quantum Physics!*, nor Irene Chan, the illustrator, has a degree in physics; however, the creation of the book was assisted by Dr. Fred Bortz, which enhances the ethos.

In their article "Using Picture Storybooks to Support Young Children's Science Learning," Rose M. Pringle and Linda L. Lamme, Professors at the University of Florida, developed a set of criteria for judging children's science books that included accurate science content; realistic illustrations; accurate "relational sizes and proportionality for illustrations" and magnification made obvious; accurate terminology; and entertaining, well-written text (4). All of these factors contribute to a general accuracy of science, which is vital for these board books to embody true pre-science.

The only informational inaccuracy in Chris Ferrie's *Newtonian Physics for Babies* and *Quantum Physics for Babies* is the use of Neils Bohr's Electron Shell Model instead of the accepted Electron Cloud Model in *Newtonian Physics*. These models describe the location and orbit of electrons around the nucleus of an atom, yet while the Electron Shell Model describes a series of rings around the nucleus in which a certain number of electrons can orbit in, the Electron Cloud Model described the uncertainty of the location of electrons in their orbit. However, this is not a significant inaccuracy considering that many students are predominantly taught the Electron Shell Model through to secondary school. It is a simplification of the atomic structure, not a meaningfully erroneous one.

The scientific ideas in Ruth Spiro's *Baby Loves Quantum Physics!* are also correct. Unlike Ferrie's books though, Spiro's book uses an analogy to simplify the quantum physics thought experiment—Schrödinger's cat. The traditional experiment suggests that if a cat is placed in a container with a decaying radioactive atom and hydrocyanic acid, which can potentially kill the cat, then one cannot determine if the cat is dead or alive until the container is opened; thus, the cat paradoxically is both dead and alive until proven otherwise. While somewhat complex, Spiro does well in minimally recasting this to make the concept understandable. Instead of the cat being dead/alive, *Baby Loves Quantum Physics!* suggests the example of the toddler's cat hiding in a box and that "until Baby looks in the box, Cat is both asleep *and* awake" (16). The concept is condensed into a simpler description, but not in a way that misrepresents the ideas. Moreover, as Patricia Lauber points out, "The writers of children's science books have always used analogies and have always looked for familiar counterparts to the unfamiliar" (9) in order to make ideas understandable, so Spiro's adjustment for the sake of the audience is largely understated.

The illustrations are also accurate. The Ferrie books use very simple ball-and-line-based images to illustrate the scientific principles, but this method is used to describe very specific factors of ideas like atomic theory and gravity, which cannot be observed. Spiro's book uses simple but fitting illustrations as well. Moreover, the Ferrie books, particularly *Quantum Physics*, do well in maintaining proportionality and clarifying magnification. Considering most of the board book focuses on the atomic level, the book must transition from a broader, familiar level (the child's ball) down to the particles. *Quantum Physics* does well in transitioning via the image of a magnifying glass over the ball showing an atom (5). On the one hand, this is a major scientific inaccuracy as a basic magnifying glass would be incapable of magnifying even close to the atomic level. On the other hand, even most adults have only a vague knowledge of electron microscopes (the instrument which can visualize atoms), so this inaccuracy is balanced with the need to signify to the target audience that atoms are too small to properly see and thus make obvious that the book is magnifying down to atoms to talk about them. Ferrie was even careful in his illustration of the atom to make the circle representing electrons to appear noticeably smaller than the protons and neutrons—something that many primary and secondary school books neglect.

Additionally, the board books ensure to use proper terminology in regards to the scientific ideas. Spiro's book only explains Schrödinger's cat and in a narrative style, so there is no need to use terminology. Ferrie's books, on the other hand, both need and successfully do incorporate scientific terminology. In *Quantum Physics*, the terms include "atoms," "neutrons," "protons," "electrons," "energy," "quantized," and "quantum." In *Newtonian Physics*, the terms include "gravity," "net force," "accelerates," "mass," "the force of gravity," and "the force of the ground."

Overall, the information is accurate, simply explained, but not distorted. This supports the value of these books as pre-science because they are not just telling a story but also realistically providing an interesting framework for scientific knowledge and thinking.

Discussion

Overall, the analysis of Chris Ferrie's *Newtonian Physics for Babies* and *Quantum Physics for Babies* and Ruth Spiro's *Baby Loves Quantum Physics!* suggests the following: (1) Based on the presumed cognitive ability of the target audience (as identified by factors such as the Kindergarten level Science Standards of Learning in Virginia), the target audience should be able to grasp some but likely not all of the concepts described in the Science for Babies books; (2) based on Apseloff's requirements for pre-literature, the books could enable the acquisition of science literacy; (3) the visual design elements are carefully selected and work to emphasize the images and concepts expressed by the books, which suggests pedagogic implications rather than mere parody; (4) the linguistic elements are designed in a way that may impart some basic conceptualization of scientific principles to the audience, but they are unlikely to be fully understood; and (5) the information is accurate, simplified, but not distorted. Thus, I argue that the role of these books is not to be amusement for adult audiences or as a truly didactic, expository book to teach children college-level physics but to act as pre-science in a future-looking pedagogy.

The books are academic in so far as they intend to arouse and sustain the innate curiosity of children by exposing young audiences to interesting and complex scientific ideas. However, the significance of the books is not in their ability to explain atoms or gravity or Schrodinger's cat to toddlers. Rather, these books are viable in their future-looking potential on a similar level to books such as the Baby Lit series to children's adaptations of Shakespearian works. As Erica Hateley explains in her book *Shakespeare in Children's Literature: Gender and Cultural Capital*, in analyzing children's Shakespeare, it is crucial to look at the relationship with the pre-text and consider the idea of "an idealized and imagined future agency. The presumption about the child as future adult … [and her] future-readership" (9). Thus, considering the pre-emptive implications of the Science for Babies series might be as crucial as analyzing the inherent design. As children's Shakespeare pushes an audience toward reading Shakespeare as an adult, these science board books prompt the rising generation toward STEM.

Works Cited

Abell, Sandra K. "Children's Literature and the Science Classroom." *Science and Children*, vol. 46, no. 3, 2008, pp. 54–55.

Apseloff, Marilyn. "Books for Babies: Learning Toys or Pre-literature?" *Children's Literature Association Quarterly*, vol. 12, no. 2, 1987, pp. 63–66.

Ferrie, Chris. *Newtonian Physics for Babies.* Illustrated by Chris Ferrie. Sourcebooks Jabberwocky, 2017.

———. *Quantum Physics for Babies.* Illustrated by Chris Ferrie. Sourcebooks Jabberwocky, 2017.

Hateley, Erica. *Shakespeare in Children's Literature: Gender and Cultural Capital.* Vol. 58. Taylor & Francis, 2010.

Hutcheon, Linda. *A Theory of Adaptation.* Routledge, 2006.

Kummerling-Meibauer, Bettina and Jörg Meibauer. "First Pictures, Early Concepts: Early Concept Books." *The Lion and the Unicorn*, vol. 29 no. 3, 2005, pp. 324–347.

Lauber, Patricia. "What Makes an Appealing and Readable Science Book?" *The Lion and the Unicorn*, vol. 6, no. 1, 1982, pp. 5–9.

Lupton, Ellen and Jennifer Cole Phillips. *Graphic Design: The New Basics.* 2nd ed., Princeton Architectural Press, 2015.

Mendlesohn, Farah. "Is There Any Such Thing as Children's Science Fiction?: A Position Piece." *The Lion and the Unicorn*, vol. 28, no. 2, 2004, pp. 284–313.

Pringle, Rose M. and Linda L. Lamme. "Using Picture Storybooks to Support Young Children's Science Learning." *Reading Horizons*, vol. 46, no. 1, 2005, pp. 1–15.

Spiro, Ruth. *Baby Loves Quantum Physics!* Illustrated by Irene Chan. Charlesbridge, 2017.

Stam, Robert. "Beyond Fidelity: The Dialogics of Adaptation." *Film adaptation*, 2000, pp. 54–76.

Virginia Board of Education. "Science Standards of Learning Kindergarten Curriculum Framework 2010." *Virginia Department of Education*, www.doe.virginia.gov/testing/sol/standards_docs/science/2010/curriculum_framewk/science-k.pdf. Accessed 12 November 2017.

Index

A. Kérchy and B. Sundmark (eds.), *Translating and Transmediating Children's Literature*, Critical Approaches to Children's Literature,
https://doi.org/10.1007/978-3-030-52527-9

The manufacturer's authorised representative in the EU is Springer
Nature Customer Service Centre GmbH, Europaplatz 3, 69115 Heidelberg,
Germany. If you have any concerns regarding our products, please
contact ProductSafety@springernature.com

Printed and bound by CPI Group (UK) Ltd, Croydon, CR0 4YY
30/09/2026
02225689-0004